This book primarily covers hobby-gr..

offroad RC vehicles. Information presented is

intended for beginners and intermediate users.

Tamiya EP models are used extensively for

demonstration.

The Tamiya models used in this book include:

Buggy Champ

Grasshopper

Hornet

Cayenne

Plasma Edge

Lunch Box

The Frog

Desert Gator

Overlander

Ford F350 High-Lift

Levant

Xtreme Truck-1

Table of Contents

End User License Agreement

This book (the "Book") is a product provided by RC PRESS o/b AirsoftPRESS (being referred to as "RCPRESS" in this document), subject to your compliance with the terms and conditions set forth below. PLEASE READ THIS DOCUMENT CAREFULLY BEFORE ACCESSING OR USING THE BOOK. BY ACCESSING OR USING THE BOOK, YOU AGREE TO BE BOUND BY THE TERMS AND CONDITIONS SET FORTH BELOW. IF YOU DO NOT WISH TO BE BOUND BY THESE TERMS AND CONDITIONS, YOU MAY NOT ACCESS OR USE THE BOOK. RCPRESS.COM MAY MODIFY THIS AGREEMENT AT ANY TIME, AND SUCH MODIFICATIONS SHALL BE EFFECTIVE IMMEDIATELY UPON POSTING OF THE MODIFIED AGREEMENT ON THE CORPORATE SITE OF RCPRESS.COM. YOU AGREE TO REVIEW THE AGREEMENT PERIODICALLY TO BE AWARE OF SUCH MODIFICATIONS AND YOUR CONTINUED ACCESS OR USE OF THE BOOK SHALL BE DEEMED YOUR CONCLUSIVE ACCEPTANCE OF THE MODIFIED AGREEMENT.

Restrictions on Alteration

You may not modify the Book or create any derivative work of the Book or its accompanying documentation. Derivative works include but are not limited to translations.

Restrictions on Copying

You may not copy any part of the Book unless formal written authorization is obtained from us.

LIMITATION OF LIABILITY

RCPRESS will not be held liable for any advice or suggestions given in this book. If the reader wants to follow a suggestion, it is at his or her own discretion. Suggestions are only offered to help.

IN NO EVENT WILL RCPRESS BE LIABLE FOR (I) ANY INCIDENTAL, CONSEQUENTIAL, OR INDIRECT DAMAGES (INCLUDING, BUT NOT LIMITED TO, DAMAGES FOR LOSS OF PROFITS, BUSINESS INTERRUPTION, LOSS OF PROGRAMS OR INFORMATION, AND THE LIKE) ARISING OUT OF THE USE OF OR INABILITY TO USE THE BOOK. EVEN IF RCPRESS OR ITS AUTHORIZED REPRESENTATIVES HAVE BEEN ADVISED OF THE POSSIBILITY OF SUCH DAMAGES, OR (II) ANY CLAIM ATTRIBUTABLE TO ERRORS, OMISSIONS, OR OTHER INACCURACIES IN THE BOOK.

You agree to indemnify, defend and hold harmless RCPRESS, its officers, directors, employees, agents, licensors, suppliers and any third party information providers to the Book from and against all losses, expenses, damages and costs, including reasonable attorneys' fees, resulting from any violation of this Agreement (including negligent or wrongful conduct) by you or any other person using the Book.

Miscellaneous.

This Agreement shall all be governed and construed in accordance with the laws of Hong Kong applicable to agreements made and to be performed in Hong Kong. You agree that any legal action or proceeding between RCPRESS and you for any purpose concerning this Agreement or the parties' obligations hereunder shall be brought exclusively in a court of competent jurisdiction sitting in Hong Kong.

Preface

R.C.PRESS is the premier information source for R/C technologies. It has the goal of putting all different kinds of R/C technologies on the global map by publishing e-books that bring to light the knowledge of R/C technology innovators.

Members of the R.C.PRESS editorial team are practicing engineers, technicians and racers who have been with R/C Modeling & Racing since the early days of Tamiya Frog and Kyosho Optima. Being geographically close to the origin of R/C products enables close contact with the major manufacturers, thus facilitating in-depth and accurate coverage of the hottest "toys of the trade".

Because we are part of the industry, we know what information is really needed, and we make sure our books tell what people really need to know. We do not mind to criticize things that don't work, and we will not hesitate to give you hacks and workarounds to difficult problems. Reading this book should be like having a R/C professional by your side, passing on useful hints whenever you get stuck.

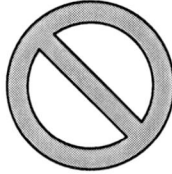

<u>Before you tackle any upgrade job, ask yourself the following questions:</u>

1. *Do you really want to do this yourself? Do understand that R/C upgrade can be either fun OR horrendous.*
2. *Do you really know how to do it? Do you have the tools to do it?*
3. *If you goof, can something be seriously damaged? And, are replacement parts readily available?*
4. *How long will it take, and how much $ will you save by doing it yourself?*
5. *In case something goes seriously wrong, is helping hand easily available?*

Proceed only if you are positive on all of the above. As a suggestion, know the locations and hours of your local hobby stores as they are most likely the closest sources of help during emergency.

Terms and theories

Terms and abbreviations used in this book that need clarifications:

- RC (or R/C): Radio Control
- rpm: revolutions per minute
- mah: milli amp hour
- diff: differential gear
- NiCD: Nickel Cadmium
- NiMH: Nickel Metal Hydride
- QC: Quality Control
- QA: Quality Assurance
- DIY: Do It Yourself
- ESC: Electric Speed Controller
- 4WD: Four-Wheel Drive
- 2WD: Two-Wheel Drive
- FWD: Front-Wheel Drive
- RWD: Rear-Wheel Drive

Tools you're going to need

The two basic types of screwdrivers are standard (slot / flat head) screwdrivers and Philips screwdrivers. Do note that using a screwdriver of the wrong type or size may damage the screw, which may give you unexpected troubles.

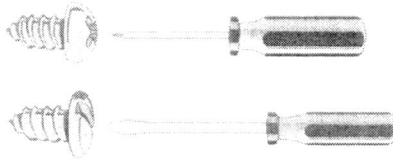

Occasionally you will come across the need to work on small screws. Having a set of small drivers like those shown below would definitely be beneficial.

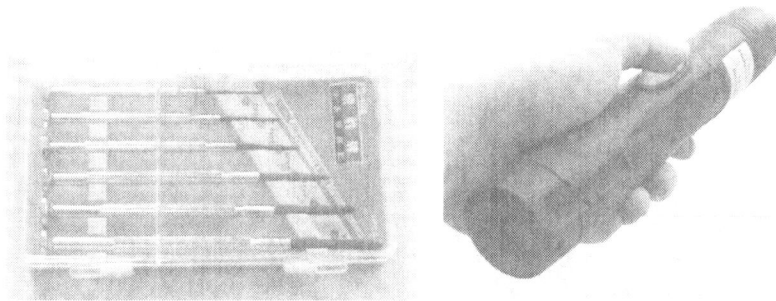

Using electric driver will definitely improve your productivity. For RC models, a simple 3.6V electric driver will do the job.

Torx screws are rarely used in RC. Torx is the trademark for a type of screw head characterized by a 6-point star-shaped pattern. Torx head screws generally resist slipping better than Phillips Head or Slot Head screws. Torx head size is typically described using the capital letter "T" followed by a number, such as T5, T10, T15 and T25. In order to manipulate Torx screws you must have the corresponding driver handy.

You may consider using screwholders for hanging onto screws that have to fit into tiny space. Many screwholders have magnets to hold the screw. Some special screwholders have a little gizmo built-in for grabbing the screw.

Screws with a centre hole that is hexagonal require the use of Allen wrenches or Hex drivers. The motor pinion gear is usually affixed via a small hex screw.

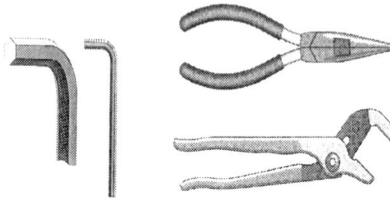

You shall need needle-nosed pliers when handling smaller screws and nuts. Combination slip-joint pliers are usually needed for handling relatively larger screws and nuts, even though they can be adjusted to several widths with a sliding pin.

You use the razor knife for task that needs a very sharp edge, such as trimming plastic and decals, cutting wire and stripping wire ends.

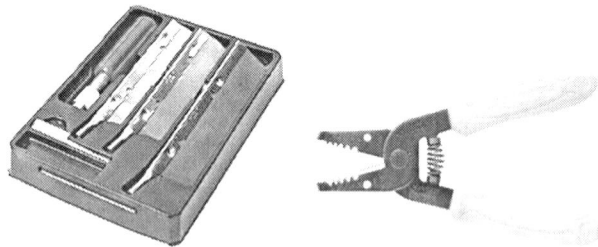

Note: If you are new to electronics or basics of wiring, don't use razor knife to stripe wires as you will likely cut into the wire strands and thus lower the total overall diameter and size of the wire, which is generally a bad thing. The best thing to do is to use a wire stripper that has two cutting edges for cutting just the insulation of a wire and not the strands. **www.home-technology-store.com** has some pretty good strippers for sale.

> NOTE: The difference between strippers is that a good one can cut both sides right through the insulation while the moderate pair can only *just* cut both sides, thus requiring you to pull the insulation with relatively greater forces (which can cause strand breakage).

You use the "E" Ring Clip Tool to effortlessly remove or insert "E" Rings (aka E clips). Some R/C cars use "E" rings a lot while some do not.

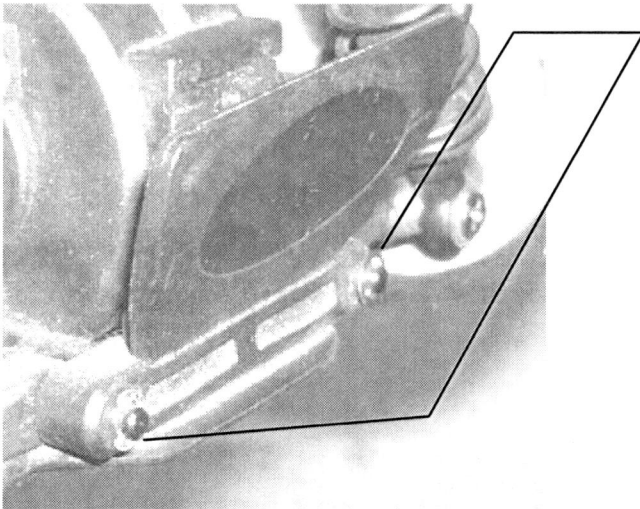

E Ring Clip Tool

I hate e-clips. They are small, easy to lose, difficult to install and remove. To prevent them from getting loose and dropping off along the ride, apply locktite!

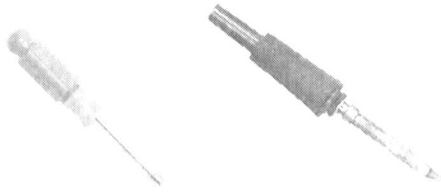

You may need to use nut drivers to set or remove hexheaded shaped screws, nuts and bolts. You may also need to use a pen shape soldering iron for rewiring the power connection.

You may find a claw hammer useful when there is a need to bang things in and prying things up. Of course, brute force is always not recommended, but sometimes you may just have no choice at all.

A drill motor (and drill bit assortment) is necessary when you need to drill holes. You want one with variable speed control as a slow setting is needed when working with ABS plastic. Most of the time a hand operated drill is good enough as you should seldom need to do heavy duty drill works on your R/C car. A drill press can help you drill better holes, but you should not need it unless you need to engage in high precision drill works. Marui used to offer an electric drill set and an electric grinding set for modeling purpose. The good thing about these products is that they will never be over powerful when working on your RC parts. The bad thing is that they require manual assembly prior to actual use :<

You need a nice place to park your tools. If you are going to work on your R/C car regularly, consider setting up a toolbox for better organizing your tools and parts. A full blown workbench with storage stack is a good idea – it lets you keep things tidy and work in a comfortable setting.

Soldering techniques

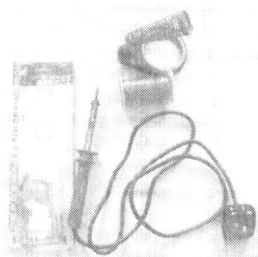

The goal of soldering is to join electrical parts together for forming an electrical connection. This is done via the use of a molten mixture of lead and tin (solder) with a soldering iron.

Supplies needed

Basic supplies needed for proper soldering include a soldering iron (the prong of metal that heats to a specific temperature through electricity), the soldering wire (an alloy of aluminum and lead), and a cleaning resin called flux that ensures the joining pieces are incredibly clean (by removing all the oxides on the surface of the metal that would interfere with the molecular bonding, allowing the solder to flow into the joint smoothly). For hobby grade usage a 4mm 60-watt soldering iron is recommended. A 20-watt one would be too weak, while a 100-watt would be way too over-powered.

Steps in soldering

NOTE:	A perfectly soldered joint should be nice and shiny looking, and should be very reliable in service. The key factors affecting the quality of the joint primarily include cleanliness, temperature, and adequate solder coverage.

The first step in soldering is cleaning the surfaces (including the iron tip itself). They must all be clean and free from contamination. Then, you may melt flux onto the parts to be joined. The parts should both be heated above the melting point of the solder but below their own melting point with the soldering iron. When touched to the joint, this precise heating can cause the solder to flow to the place of highest temperature and makes a chemical bond. Do keep in mind that too much solder is an unnecessary waste but too little may be insufficient for supporting the component properly.

Warning: !	Solder melts at around 190 degrees Centigrade. Such a temperature is hot enough to inflict a nasty burn. Be extremely careful when you do your soldering work. Also, do your soldering work in a room with good air circulation. Soldering does release toxic fumes.

Soldering battery cells

Soldering directly to the battery cells is never a good idea. You should only solder to the welded tabs. Soldering to the cell button can destroy the nylon seal. You simply can't get the cell button hot enough to get a good solder joint without compromising the integrity of the nylon seal ring. Warning: this does not apply to Lipo/Life cells. Never manipulate these cells on your own!

Proper handling of the screws

Most amateurs tend to screw things very tightly in hopes that the parts will not fly off later. This is in fact a deadly mistake because some screws, bolts and nuts are NOT supposed to be tightened too securely or the threads would be stripped. If you find yourself confronted with a screw that is extremely difficult to get unscrewed, don't use brute force (or you risk stripping the threads). Instead, give the screw a slight twist in the opposite direction and then loose it again. If this does not help, tap the screwdriver on the head with a small hammer (but don't tap it too hard). If it still fails to make it, try to squirt the screw with penetrating oil or WD-40 and retry. Use WD-40/RP7 with caution. WD-40/RP7 has a corrosive nature and is generally not recommended for use on plastic parts. To be safe, consider using silicon oil spray instead.

Silicon oil, safer for models with plastic parts

WD40/RP7

Stripping screws can be frustrating. If unfortunately you strip a screw, the 2 easy ways to remove it are:

- Fill the stripped screw hole with J.B. Weld (which is a type of glue specially for use with metal parts), and then put your screwdriver into the old hole to create a new fitting. Give it 10 to15 minutes to set and dry completely, then unscrew it.

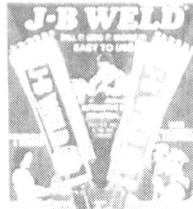

- Drill a hole in the screw, then scoop out the old screw.

On the other hand, to prevent certain critical screws from getting loose, apply threadlock/locktite - a glue type of compound that makes screws more secure. Loctite usually won't bite into plastic very well. It can sometimes soften the plastic, but most of the time it won't really be permanently stuck on there. Most of the time you can get a locked screw loose via the use of a decent screwdriver (by the way, heat is what is used to release excessively strong locktite).

NOTE: Anywhere screws thread into metal, apply a dab of locktite for preventing all the vibration from loosening the screws.

There are locktites of different strength available. You may want to check out the proper type to use through this URL: http://www.loctiteproducts.com/glue.asp

Generally speaking, there is no need to use locktite on a brand new R/C car unless the car is going to run entirely on bumpy tracks (meaning the car is going to shake excessively all the time).

Frequent disassembly can make things loose, therefore you may need to apply locktite accordingly. It is especially suggested that you use locktite on critical parts like the nut that holds the wheel in place.

If there are LOTS of screws to work with, one good thing to do is to take out a piece of paper, draw a rough sketch of the pieces you are taking the screw out of, then get some tape and roll it so that you have a sticky side on both the top and the bottom. Draw circles on the sketch and place the tape/screw inside the circle. This is very useful when you are taking something apart that has more then say 15 screws.

The proper ways of taking things apart and putting them back together

To avoid getting into chaotic situation when doing your assembly and disassembly works, try to be as organized as possible. Do your work on a clean, dry and flat surface which is close enough to reach without having to walk back and forth for grabbing the necessary tools. *In fact, it is best to work things out on a big white towel. The towel provides a color contrast, thus making it easy to see the parts as you lay them down.*

Before you remove each part, ask yourself the following questions and take notes if possible:

1. What is this, and what is it for?
2. Why is it made the way it is?
3. How tight has it been screwed on out-of-the-box? How hard is it to remove?

As you remove each part, lay it down on a clean flat surface in clockwise order, with each part pointing in the direction it laid when it was in place. Assign each part a number indicating the order in which it was removed. When you are ready to put them back together, start with the last part you removed and then go counterclockwise through the rest of the parts. If possible, go to your manufacturer's website (or any other web site) and look for any documentation files (mostly in PDF format) they offer for free download. Many of them include very detailed fly-out diagrams, a complete list of all parts as well as where they fit. This will greatly lessen any confusion you might have when putting things back together.

Warning: !	Along the process of disassembling/assembling your R/C car, there are chances for objects to fly off (especially when springs are involved). Therefore: • keep your children and pets away from the work area. • don't work in a location too close to the windows (you don't wanna scratch the window glass).

Your Upgrade Objectives

Friends in the US ever heard of the rumor that the 55 miles-per-hour freeway speed limit was set partly for energy conservation? At this speed level, maximum fuel economy and engine lifespan can be achieved. If you go faster, however, the gas mileage will become un-proportionately poorer and your engine will require maintenance work much sooner.

In the world of R/C, it is possible to push the limit to well over the factory default level. The thing is, there are always trade-offs between energy consumption, lifespan, power and $$$. If you have load of $ to spend then you don't need to read this book at all – just ask for the most expensive upgrades and you're ready to rock n' roll. If, however, you want to spend less and keep cost-effectiveness in mind, then it is recommended that you first figure out your upgrade objectives and the associated trade-offs before making a decision.

Possible upgrade objectives:

Controllability – You want a controllable ride (the ability to make sharp and responsive turns)?
Speed & Stability – You want a speedier and more stable ride (straight line stability)?
Endurance – You want a longer lasting ride (energy consumption)?
Sturdiness – You want a crash-proof body?

To help you understand the achievability of these often contradicting objectives, let's look at the structure of an electric R/C car:

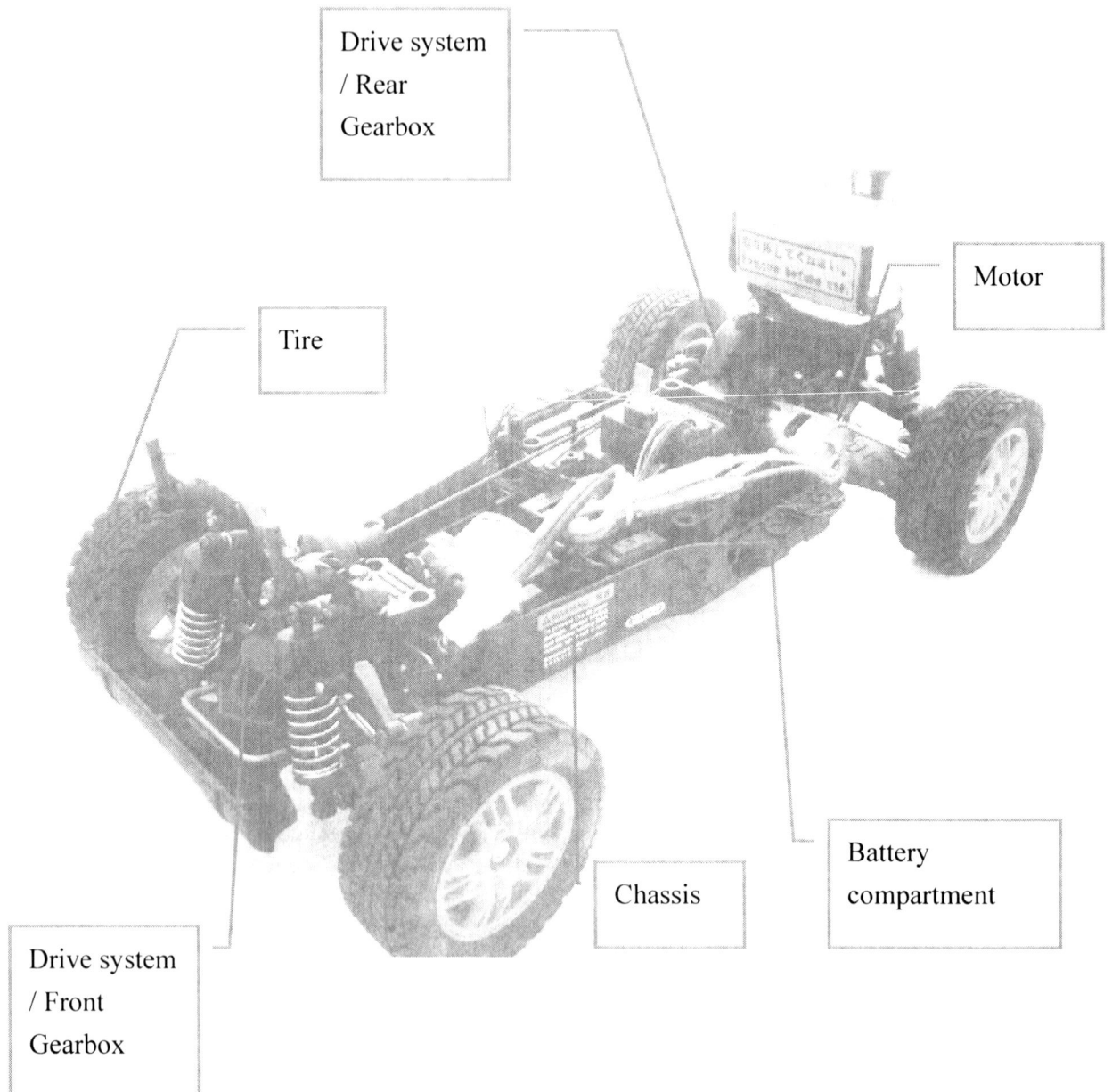

Drive system / Rear Gearbox

Motor

Tire

Drive system / Front Gearbox

Chassis

Battery compartment

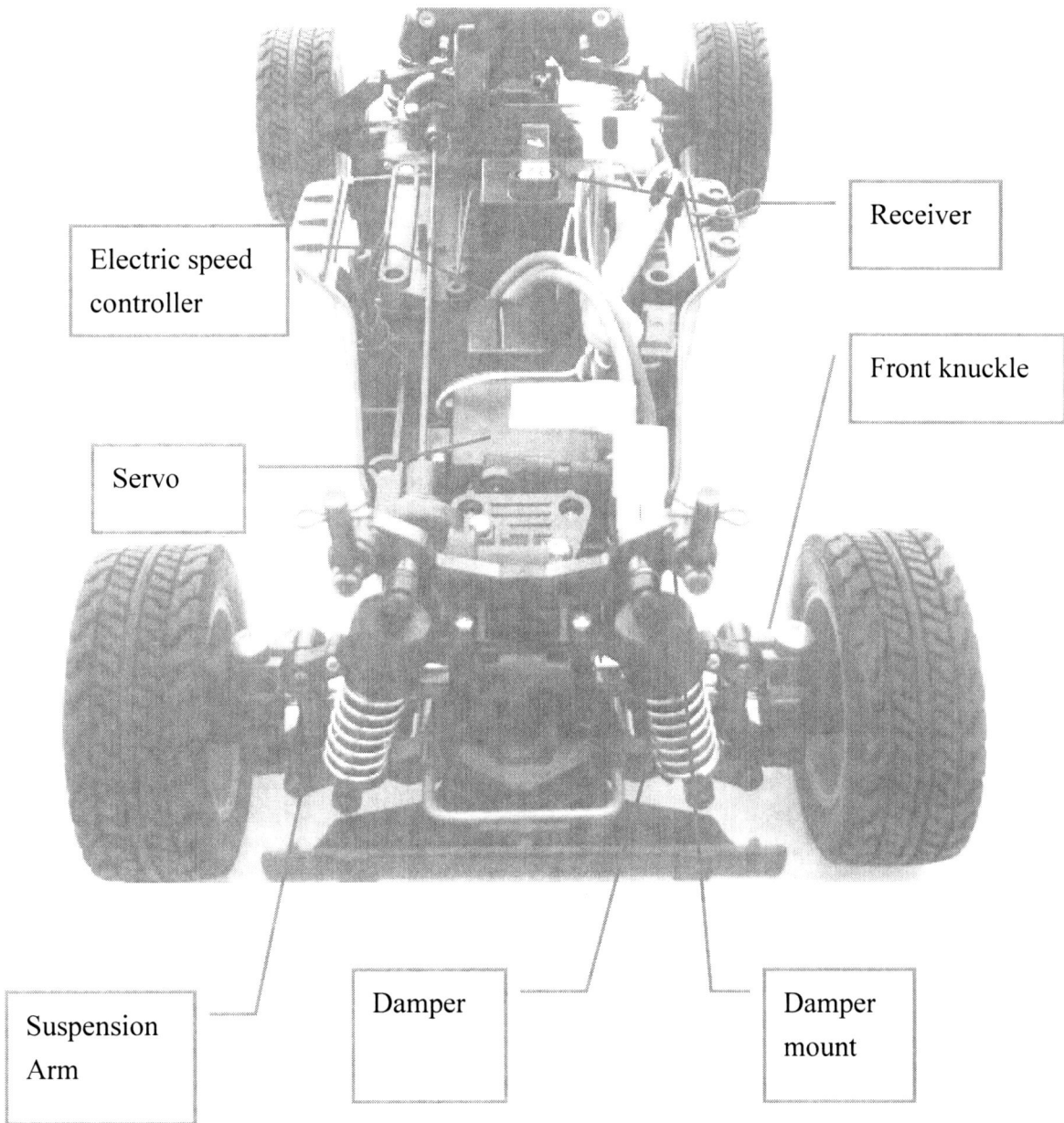

Electric speed controller

Receiver

Front knuckle

Servo

Suspension Arm

Damper

Damper mount

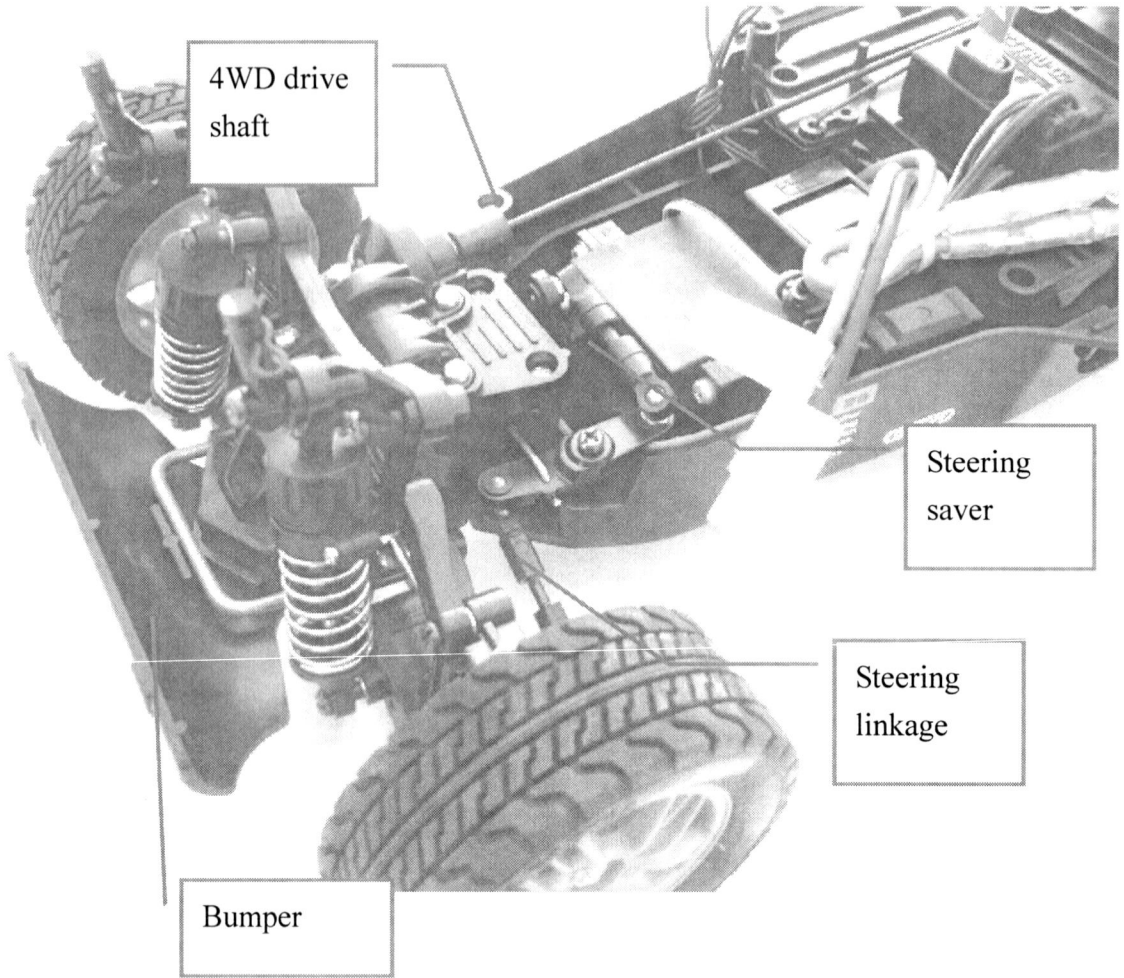

4WD drive shaft

Steering saver

Steering linkage

Bumper

Rear hub carrier

Dog bone shaft

The pictures show the critical components of an electric R/C car. Each of these components has a role in achieving your upgrade objectives. We will deal with them one by one throughout the rest of this book.

NOTE: In the real world, every bit of change on each of those critical components can make a difference towards the car's overall performance. Since budget is always limited, you just can't upgrade everything. Therefore, you must properly prioritize your objectives and goals with an attempt to balance the tradeoffs involved:

- There are tradeoffs between the contradicting objectives of Controllability and Speed/Stability. The way you adjust the suspension system and the tires can affect them most.
- Tuning of the drive system (such as the gear ratio), the motor and the battery can have significant impact on the contradicting goals of top speed, acceleration and endurance.

Also keep in mind that different people have different preferences on their upgrade strategies. Since so many factors come into play, there is simply no single correct answer to the question of what upgrade does best. We give you expert advices based on our many years of electric R/C experience. However, there are always better ways of doing things. Therefore, please feel free to incorporate our advices with your own findings and tailor-make your own unique R/C strategy – one that truly fits your own practical needs.

Assembly Kit VS Ready-To-Run

If what you have is a kit which requires assembly, you may plan for parts replacement early and skip the steps of taking things apart. The good thing is that you can save time – you don't have to take things apart before putting in your upgrade parts. The bad thing is that you don't have the chance to try things out prior to making the upgrade decision – you may spend $ in the wrong places. For beginners, building the entire car from scratch may be a little too difficult anyway.

RTR (Ready-to-run) kit may be a better choice for beginner since the car is built to be ready to go. You may start with the stock configuration, drive it for several times and learn how it works. As you gain experience, you may get your hands on the more advanced stuff like fine tuning and basic disassembly.

The Overlander XB (RTR) 2WD.

The Lunchbox XB (RTR) 2WD.

The Plasma Edge XB (RTR) 4WD.

With a RTR package, everything is in place so the first thing you would want to do is to check and ensure the radios and the battery are working. Charge the battery using the charger that comes with the package, then hook it up with the ESC and test out all the remote control functions. Then check the differentials. When you spin one wheel, the other one should spin in an opposite direction. Also check the screws. Make sure all screws are properly tightened. Before test driving it on the road, lube the spinning parts and joints that are associated with the four wheels with silicon oil spray. Don't forget to also lube the motor bushing. For an initial upgrade (brushed configuration), you may want to install a 23-T Sports Tuned motor (not for crawler though), which is considered pretty safe for the stock ESC and the stock battery. If you want to upgrade the battery, instead of pumping up to 8.4V you may want to go for a high capacity 7.2V pack (something over 3300mah) OR a 7.4V Lipo pack such as the Speed Energy 4200mah pack, which will for sure give you more horse power and longer running time, without the risk of cooking the stock ESC.

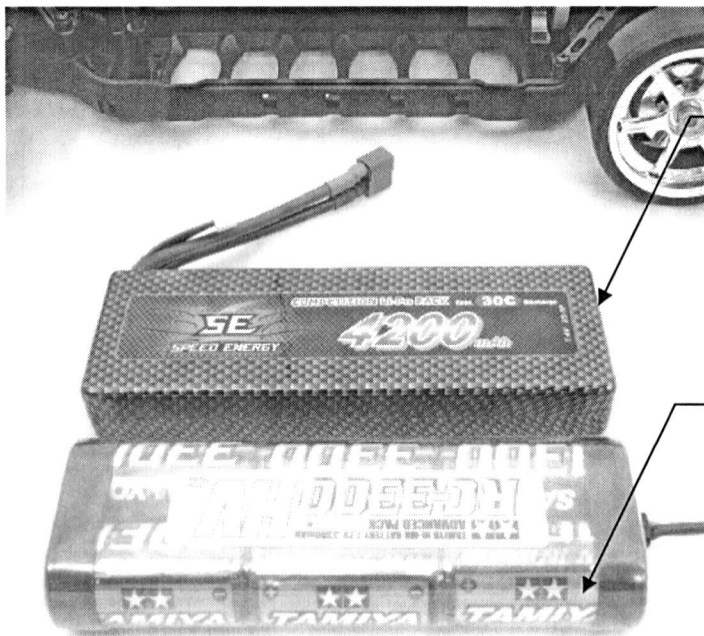

The Speed Energy 4200 mah Lipo pack is a good choice for RC.

A typical 3300mah pack can provide approx. 10 to 15 minutes of ride time.

Copyright 2010. **The R.C.PRESS (Hong Kong)**.

<table>
<tr><td>NOTE:</td><td>The ESC that comes with a RTR kit may not be able to handle power level that is way higher than that of the stock configuration (the BEC unit may be damaged). Therefore, do NOT attempt to run a 7.2V RTR car with a 8.4V pack unless your vendor confirmed that it is safe to do so.</td></tr>
</table>

The stock ESC is never for serious racing:

The chassis and its material

When you shop for a new R/C car, one primary consideration is the chassis. Factors that deserve your attention include weight (and weight distribution), sturdiness, and battery compartment. Plastic (nylon) boards are heavier but are flexible enough to stand up to abuse. Graphite boards are stiffer and lighter but are (relatively) more brittle (as they are less flexible). Plastic (ABS) bodies are very popular among the lower end Tamiya RC models. They are not too heavy but are quite brittle too. Metal bodies are very solid and are most ideal when a lot of bashing is expected.

Bath tube style (tray like) chassis is made of ABS plastic.

A crawler doesn't really have a solid chassis body. It is more like a set of skeleton.

Metal chassis material is very strong but is also very heavy.

Weight & Weight Distribution

A sturdier structure usually features reinforcement parts throughout the body and is therefore heavier. Fortunately, there are many ways for "slimming" your car. Basically, on a chassis like the one shown below, you may drill holes on areas which do not affect the overall sturdiness of the chassis platform (of course you don't want a car full of holes on the chassis to run on dirty tracks):

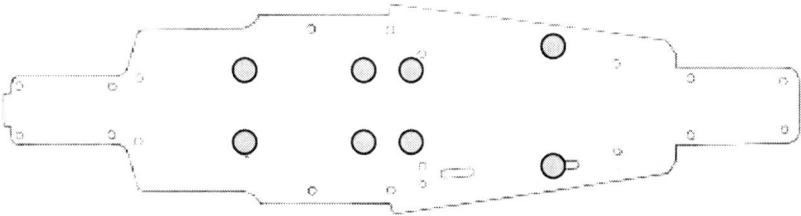

Note: drilling may be done on ABS chassis or graphite chassis with caution (don't press too hard too sudden or the chassis will break)

Most chassis designs do not give you the freedom of moving things around at will. You

Copyright 2010. **The R.C.PRESS (Hong Kong)**. All rights reserved.

may at the most swap the positions of the receiver and the ESC (and perhaps some battery cells). Also, the more components you load on the chassis, the lower ride height you are going to have. A low ride height tends to give better stability and is not preferable when lots of jumping actions are expected. **For crawler: a low center of gravity is preferred when dealing with a steep incline. A center of gravity that is too high will make rollover fairly easy.**

Due to the fact that most R/C cars have their motor placed closer to the backend of the chassis, the rear and front ride heights tend to be uneven (the heavier side goes lower). An artificially configured "lower front" can benefit steering but is easier to go "diving" (which can make the car slides) when you brake real hard.

The TL-01B is a shaft-driven 4WD chassis which has a rugged clamshell design with long suspension arms and long dampers. The DF-02 chassis is a shaft-driven 4WD chassis with a longitudinally mounted motor and battery placed in a wide bathtub design. The DF-03 chassis is a shaft-driven 4WD chassis with a monocoque bathtub frame.

For RWD in particular, the motor is usually located closer to the rear end.

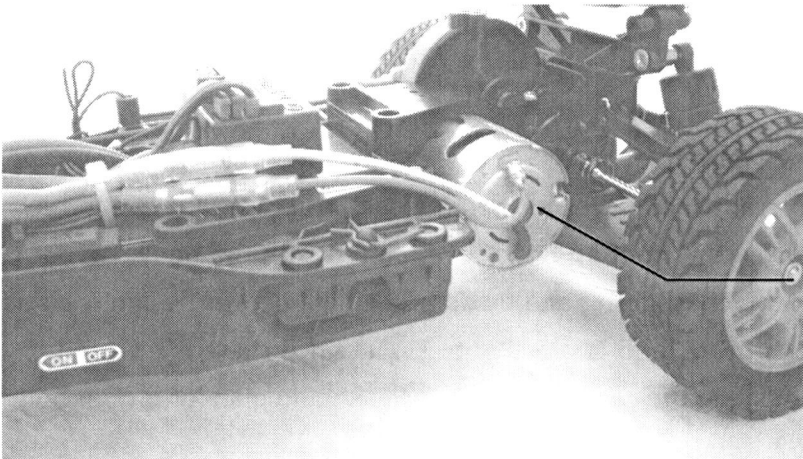

For 4WD, the motor is usually located a little closer to the middle.

As said before, there isn't much you can do to redistribute the body weight through rearranging the component locations. Therefore, should imbalance of weight be found (or should you want to have an intentional uneven ride height setting), all you can do is to adjust the front and rear suspension systems accordingly.

Due to space constraint, there isn't much flexibility in moving parts around.

Copyright 2010. **The R.C.PRESS (Hong Kong)**. All rights reserved.

Structural sturdiness

The sturdiness of the chassis largely depends on the chassis layout. Refer to the illustrations in the next several pages, layout A has a weak structure that can hardly survive a crash due to the way the front and back are jointed with the body. Layout C has a double decker design which is strong and flexible. A single decker design such as that of layout D is less crash-proof, and is mostly for RWD (since there is no need to worry about 4WD shaft damage in the case of a crash). You should plan for proper body reinforcement accordingly if frequent crashing is expected.

Do note that body structure reinforcement often involves the introduction of extra tailor-made parts (or the replacement of certain stock parts with something more heavy and solid), therefore weight increase (together with poorer energy efficiency and loss of speed) is hardly avoidable.

The ideal choice of reinforcement material (if you are going to tailor make reinforcement parts) is graphite. It is strong, light and easy to cut. However, it is relatively expensive and may not be readily available (3Racing does have graphite reinforcement parts for certain Tamiya models). Metal is rock solid but is heavy. Strong plastic is the weakest among the three but is cheap and widely available. Since we can always come up with ways to slim the chassis, a little bit of extra weight brought by a piece or two of strong plastic shouldn't hurt much.

Stadium Trucks are primarily designed for jumps and off-road racing, while Short Course Trucks are made to be extremely durable (to absorb contact from other vehicles better than traditional RC trucks) for frequent bashing. As of the time of this writing I do not see any Tamiya Short Course Truck available.

Stadium truck has a small jumper due to concern on weight and weight distribution. There is also no side-protection.

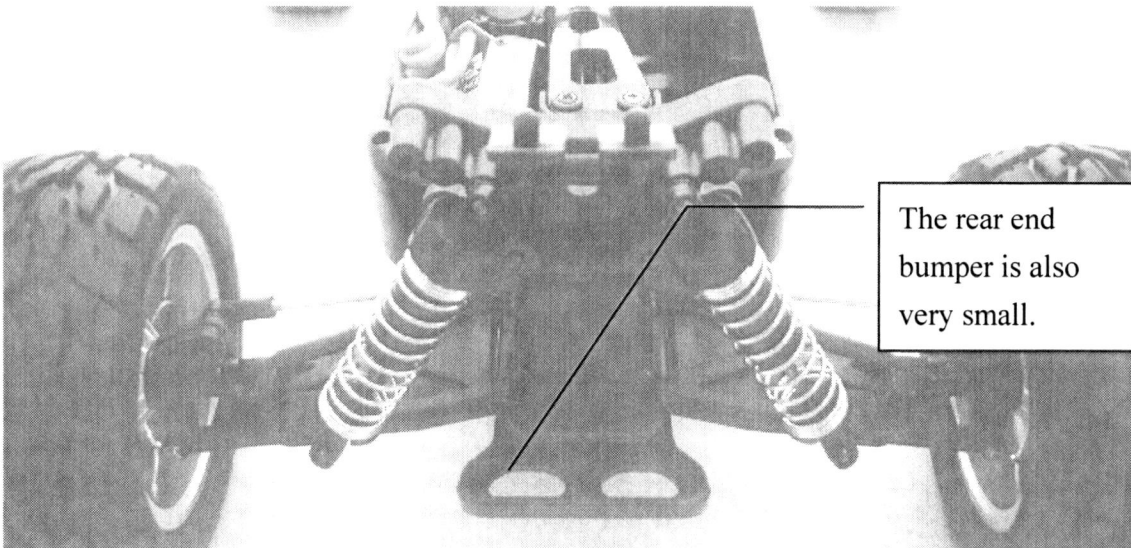

The rear end bumper is also very small.

Copyright 2010.   **The R.C.PRESS (Hong Kong)**.

Layout A:

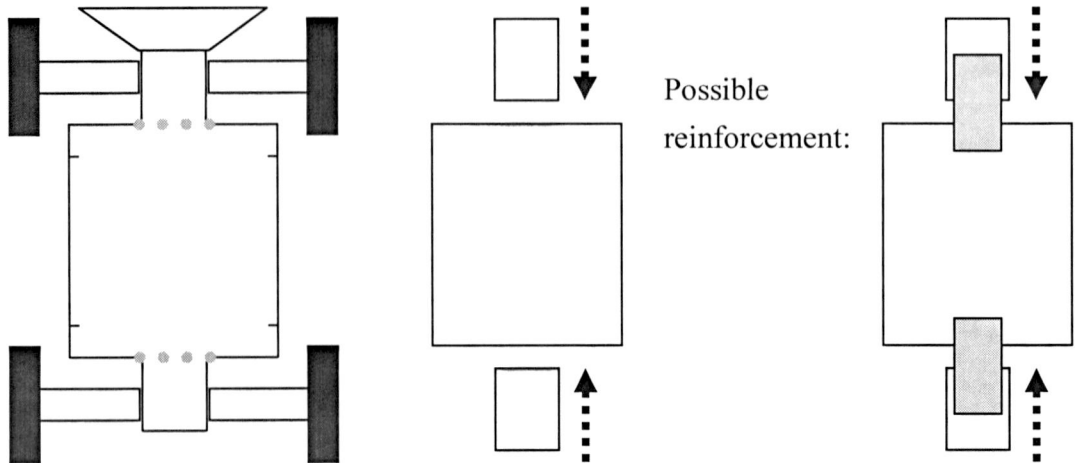

Possible reinforcement:

The older generation Tamiya Falcon has this type of chassis layout. The front intersection is very easy to break upon a crash. A limited number of modern RWDs still deploy similar designs.

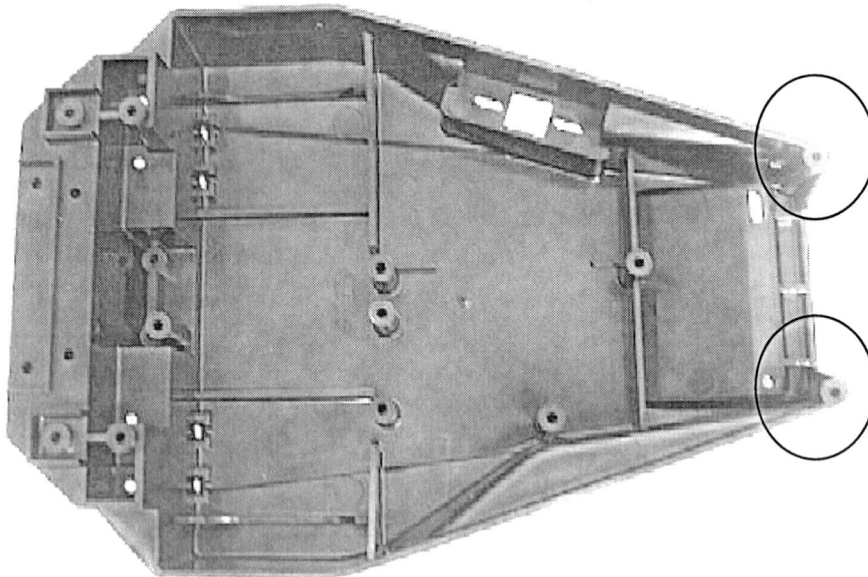

Layout B:

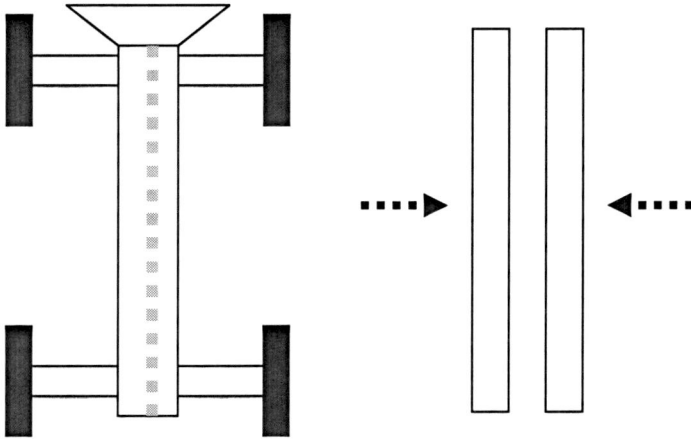

This type of chassis is pretty solid so the need for reinforcement is minimal. Overlander, Dual Hunter, and Twin Detonator are based on this chassis design.

Layout C:

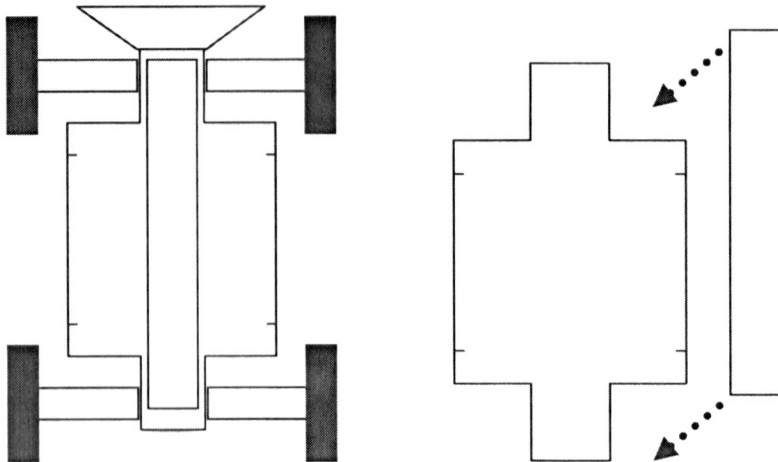

A double-decker chassis layout is preferable for 4WD cars. It offers sturdiness, simplicity and flexibility by design. It also offers better protection of the shaft during a crash.

Layout D:

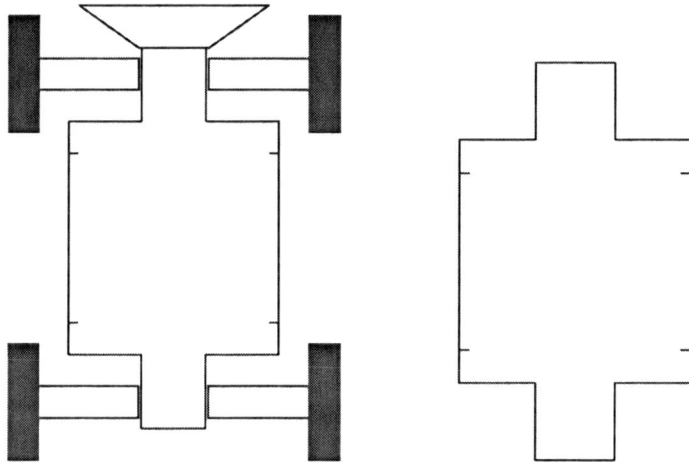

A single-decker design is weak. Regardless of how hard the chassis material is, the chassis is going to bend upon a crash. That is why it is primarily used in RWD (so no need to worry about shaft damage). To properly reinforce it, the best thing to do is to add an upper deck to the design.

This piece of plastic is NOT an upper deck but just a battery cover.

If there is not enough space to accommodate an extra deck, you may want to install two tailor-made metal rods on the sides instead (just to ensure that they won't get in the way of the belt/shaft and the steering mechanism):

Possible reinforcements:

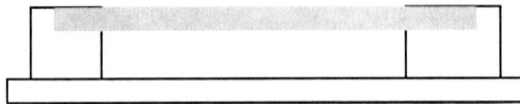

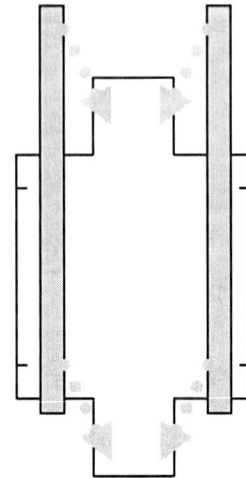

OR

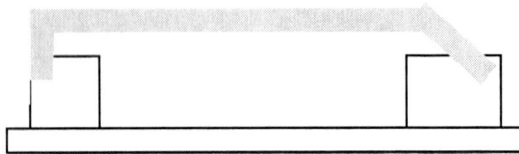

OR

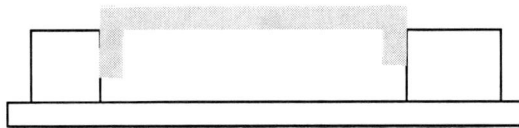

OR:

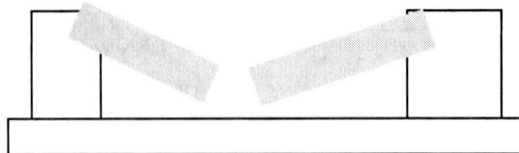

Crawler has a skeleton like structure. Since it is never intended for bashing, crash protection is not necessary. Design focus is on flexibility – the ability to twist.

Battery compartment

The way your battery is housed depends on the chassis layout. With a chassis like the one shown above, the battery compartment is not flexible enough to allow for anything larger than a standard 7.2V pack. If you want to use a 8.4V pack, you'll need to place an extra 1.2V cell somewhere else (see the picture below) or go with something smaller (such as the GP 1100mah mini cells – the problem with smaller cells is that they may just be too weak to push a sport tuned motor).

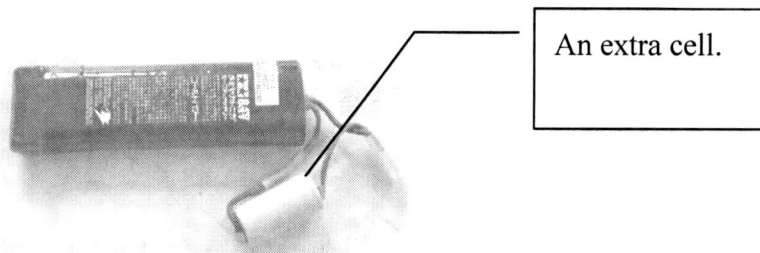

An extra cell.

Manual modification necessary to "extend" the battery compartment.

Regardless of how you wire your cells, the key is to keep the body weight properly balanced. NEVER attempt to compensate for any imbalance through adjusting the suspension on only one side of the car (the idea is simple - if you adjust the left, you must also adjust the right. Both sides should always go hand in hand).

Scale is another factor that must be considered for arranging the battery layout. For smaller cars you need battery of different sizing:

The 1:12 Lunch box has a pretty small battery compartment.

Bumper

When there is a crash, the most vulnerable parts at the front are:

- the front lower arms
- the front upper arms
- the steering rods
- the front drive shafts (FWD or 4WD)

Most other R/C cars are equipped with small bumpers due to the need for weight reduction (and for fitting with the plastic body shell). One best way to protect these parts is to extend (widen) the bumper. Refer to the following pictures, a small bumper like this is not capable of protecting the entire front arm assemblies:

This stock bumper is a bit small.

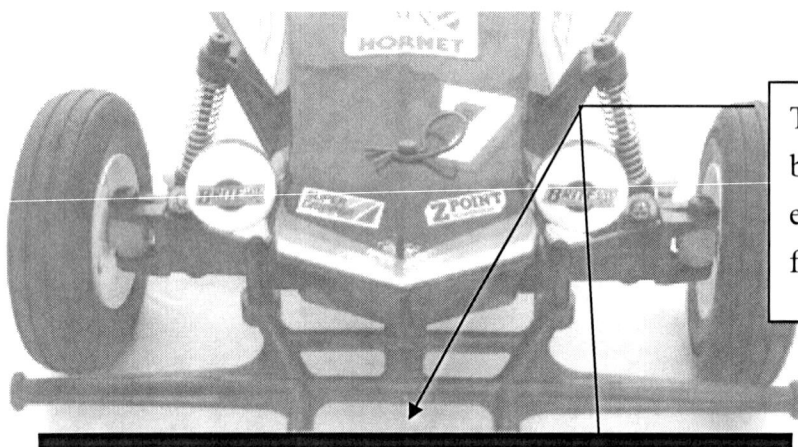

These stock bumpers are wide enough to offer full protection.

By properly "widening" the bumper using some light weight material, better protection can be achieved. Since the primary purpose of the bumper is impact absorption, you may use sponge as the choice of bumper extension material. Sponge can absorb impact effectively without adding much weight to the chassis.

Do note that pro racers always dislike the idea of large bumper due to concern on weight. For beginners, however, a large bumper can protect their valuable first time investment, at the expense of slight weight increase (which wouldn't matter much for them anyway).

Short course trucks are always equipped with large extended front/rear bumpers. For stadium truck, if the bumper fails to provide adequate protection, you may need to replace the stock plastic front arms (most come stock with ABS plastic arms) with something stronger, such as those made with Aluminum Alloy (they are in general heavier than the stock plastic arms though).

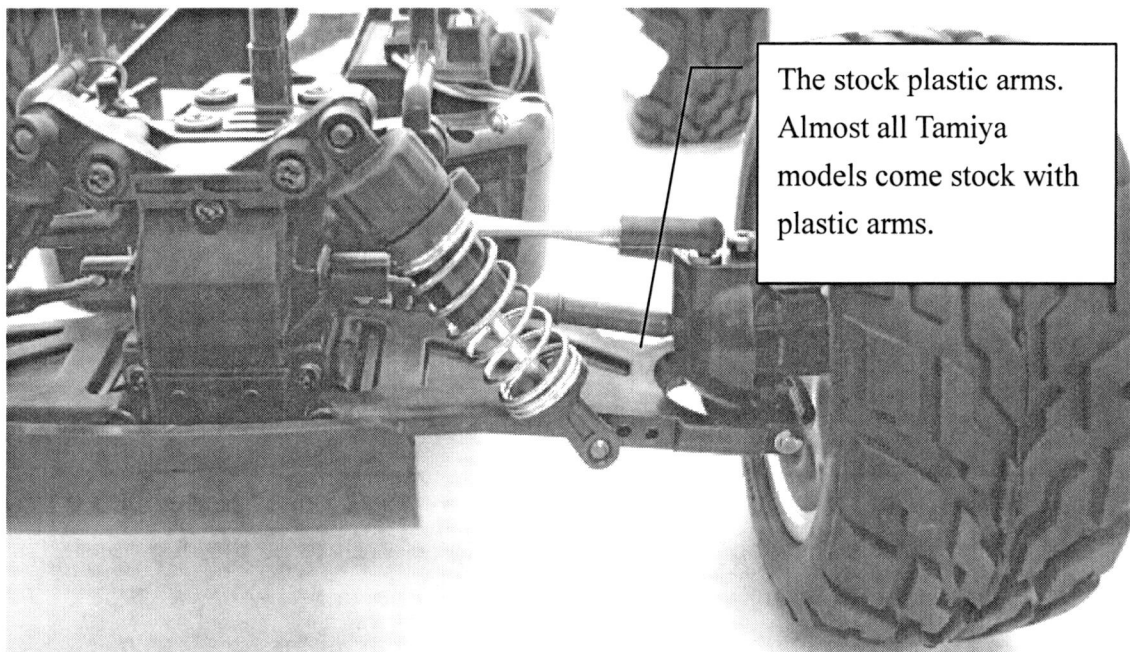

The stock plastic arms. Almost all Tamiya models come stock with plastic arms.

Copyright 2010. **The R.C.PRESS (Hong Kong)**. All rights reserved.

You should also consider replacing the stock front drive shafts with a pair of titanium shafts. The stock shafts are made with steal and are very easy to get bent upon crashing. Titanium shafts are in general lighter and much harder. Also, the piece that holds together the arms should be made with stiffer metallic material. Plastic is not preferred here.

Strong aluminum/metal brace preferred at the front.

Strong aluminum brace also preferred at the back.

NOTE: Note that the traditional type of drive shaft kinda reassembles a dog's chew bone and is often being referred to as the dog bone shaft. An enhanced type like the one shown below has a pivot type joint on one end and is often used in professional racing under a ball diff configuration.

Another worry is on the side – when a car goes sideway, a sideway crash can occur. For beginners in particular it is important to have adequate protection on both sides of the chassis. One easy way is to have foam strip attached to both sides of the body shell. The foam can absorb some impact during a crash. You don't need to do this for short course trucks since by design they all come with adequate side protection.

Side protection is usually not sufficient in a stadium truck.

The drive system

Traction

To better understand the drive system it is necessary to know what traction is all about.

Traction refers to the resistance or friction happening between each tire and the ground surface. More traction allows your car to turn at higher speed without losing control.

To transfer the motor power into moving force, an equal amount of traction is needed. Technically speaking, traction is the result of the following factors combined:

- the friction between the tire and the ground
- the area of ground covered by the tire
- the car weight which presses the tire onto the ground (in other words, weight contributes positively to traction)

2WD VS 4WD

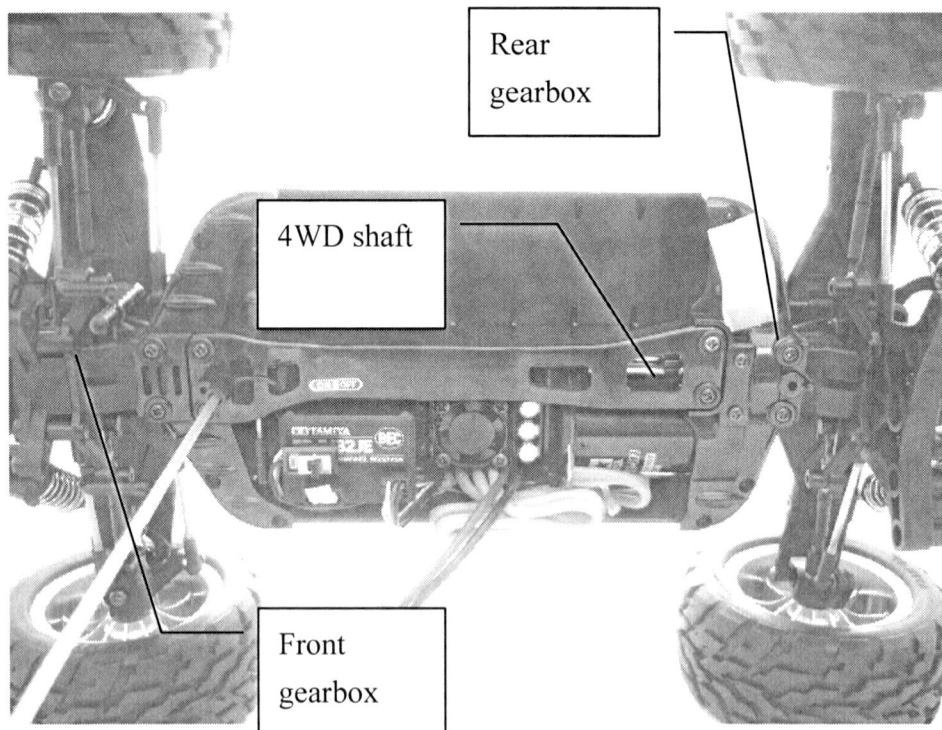

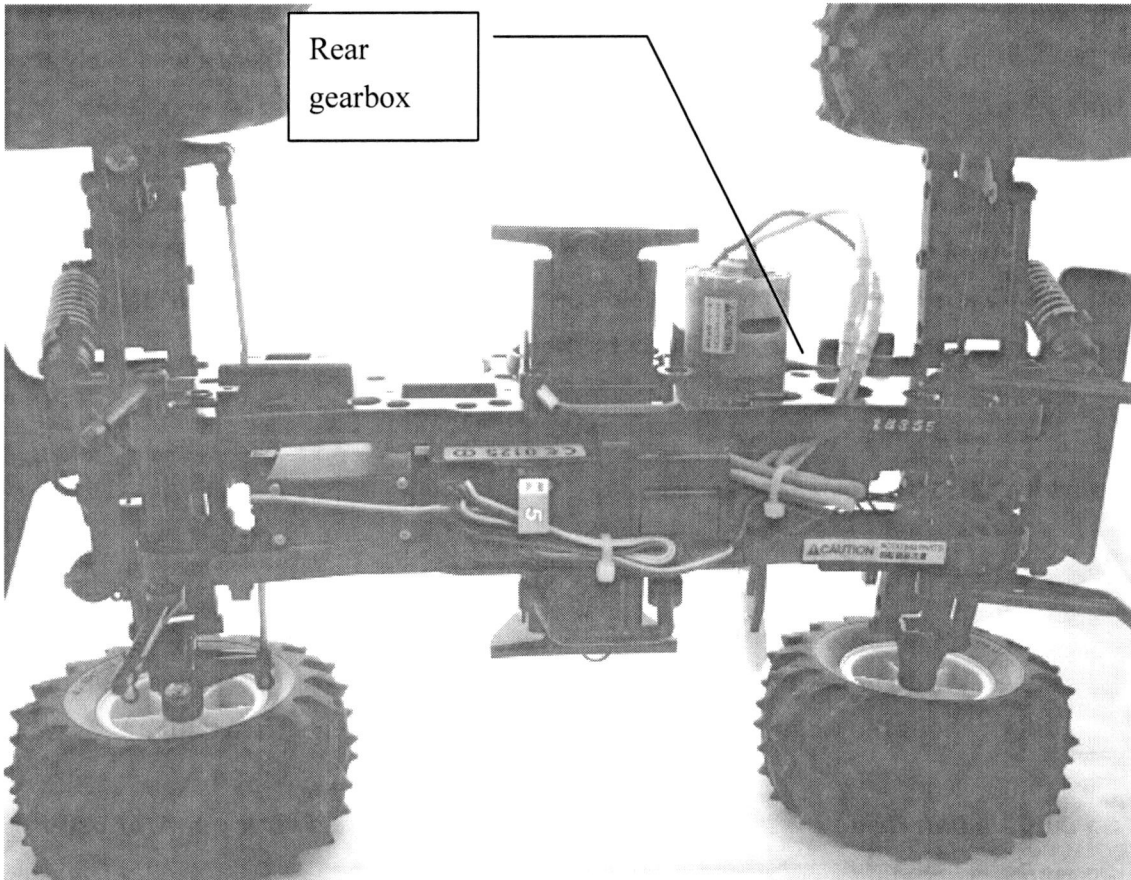

Rear gearbox

In 2WD (FWD and RWD), the traction of only two wheels is used (the other two are just rolling along). When more torque is applied than there is traction available (you accelerate too sudden and too much), the two powered wheels will break lose and start spinning. When the tires are spinning, your car may start to lose traction (and you may start to lose control over the car). It is believed that RWD is even worse than FWD in this regard.

So, if your car needs more torque to accelerate (you need more torque for acceleration. Once you accelerated to a desired speed level, you need more rpm to maintain that speed), more accompanying traction must be available in order to avoid slipping wheels and maintain control over the car. In 2WD, the entire driving force is directed towards only two tires. In 4WD, the driving force is "shared" among four tires. Since each tire in a 4WD setting has to support a much smaller torque load, the tires are less likely to break lose.

Try it out on a road with poor traction (such as on a sandy road). When accelerating a 2WD, you must do it gradually bit by bit or the car will immediately spin and lose control. With a 4WD it is much easier to maintain control. **In the world of real vehicle short course/stadium truck racing, RWD is the mainstream.**

Shaft drive system VS belt drive system

There have been debates as to which drive system is better than the other. Frankly speaking, there are pros and cons for each, and it is strictly a matter of personal preference. Belt drive has a simpler design due to the use of less gears along the drive chain, and it keeps the weight down for the same reason. It tends to run quieter as well.

The DB01's belt drive system features a centrally-located spur gear for driving two belts.

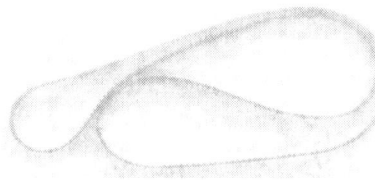

However, the belt is more subject to wear and tear (it will start weakening and stretching over time) especially upon frequent hard accelerations and you will need to regularly adjust its tension just to prevent it from skipping. Also, it is widely believed that belt drive car often induces drag (i.e. belt drag), which can make the drive system quite inefficient. An improperly sealed belt drive system is another source of problem –

it can get stuck easily if dust and unknown debris find their way into it.

Dust can get in quite easily. The way the belt drive system is structured makes it quite difficult to effectively seal the system.

The new TA05 chassis from Tamiya comes with a newly designed transmission system which employs two equal-length drive belts mounted front and rear (interlocking with tooth surfaced center pulley) for reducing drive resistance and improving transmission efficiency at high speeds. However, this more complicated setup makes effective sealing of the belt drive system even more difficult.

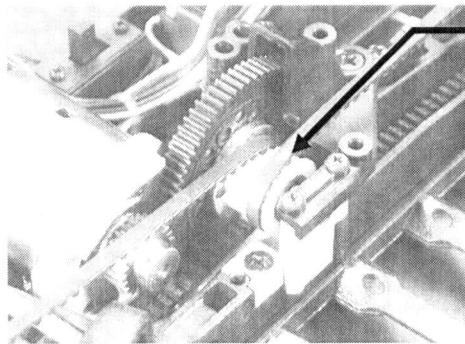

A system with 2 belts.

Shaft drive gives better response and controllable power on demand. As long as the drive system is properly sealed, it is hassle free and almost maintenance free (it is much more difficult to break a shaft than a belt after all). **Almost no tuning is required on the shaft.** However, a shaft drive system is heavier (due to the use of more gears) and is usually much louder during acceleration. Also, if the chassis flexes then the gears may be bound by the shaft, which could lock up the tires and "jam" the car entirely (a belt, on the other land, gives "flexibility" through "skipping").

Aluminum shaft is light-weight and sturdy (if you don't crash your car real hard).

NOTE: It is technically possible to convert a shaft drive car into a belt drive car (and vice versa), but the cost involved may be way too unjustifiable.

A very light -weight thin shaft

For rock crawler and very high lift truck, the motor is always at the center and there are two independent shafts for connecting to the front and rear gearboxes.

Expert tips

- If you are going to do lots of off road racing on dirty tracks, go for a shaft driven 4WD. There is practically no perfect way to sufficiently seal a belt drive system unless the system was designed to be sealed from the ground up (the vintage PB Mini Mustang has a good chassis design which effectively protects the entire belt drive system. This kind of design can be rarely found nowadays).

- If you are in an area with extreme weather conditions, go for a shaft driven 4WD. The rubber material used by some belts is very sensitive to differing weather conditions (too cold VS too hot) and may require frequent tension re-adjustment.

- If you insist on getting a belt drive car, be sure to do whatever you can to seal the belt drive system. Some belt driven 4WD packages come with a set of transparent plastic cover for protecting the belt (like the one shown below). If possible attach a stripe of sponge tape on each side of the cover to prevent dust from getting in.

Gears and bearings

Gear ratio and gear pitch

In its simplest form, the term "gear ratio" defines the relationship between the pinion gear and the spur gear. It addresses the concern of how many times the pinion gear has to rotate in order to make the spur gear turn around once. Therefore, gear ratio equals the number of teeth on the spur gear divided by the number of teeth on the pinion gear.

To illustrate, refer to the example illustration below:

The above shows a 3:1 gear ratio. The gear with 10 teeth (the right one) will have rotated completely when the 30-tooth gear (the left one) has rotated a third of its way.

The "gearing" process actually reduces speed, not increases it. Such a speed reduction is necessary as it adds torque. Keep in mind:

- A low gear ratio slows things down but produces more torque.
- The higher the gear ratio, the hotter the batteries and the motor get (heat is no good as it increases energy consumption) as they induce more current to supply the same amount of torque (but you can get a higher top speed - a gear ratio closer to 1:1 will allow both gears to turn at closer speed rates, thus producing higher rpm).

Simply put, on the same motor less torque means:

1, higher top speed.

2, more stress imposed on the motor and the battery when going uphill.

3, more heat and higher energy consumption.

On the other hand, more torque means more power and better acceleration but less top speed. Tire size does make a difference. If you are running on larger tire (like a truck), more horsepower would be required and you would want to use a low gear ratio. If you are running on smaller tires (like a touring car) you need speed and higher gear ratios would be beneficial.

The legendary PB Mini Mustang has a dual gearset design (2 sets of pinion/spur gears and a clutch) which allows for automatic gear shifting. No other manufacturer has ever followed suit though.

NOTE: When browsing through the R/C product brochures, you often see the term "GearBox Ratio" which says something like 8:1 or 7.5:1 …etc. As an example, the Kyosho PureTen EP Alpha 2 4WD has a factory default gearbox ratio of 8:1, meaning the motor has to complete 8 rotations in order to turn the wheels around just once.

Comparing 8:1 with, let's say, 10:1, 8:1 gives less torque but higher top speed. 8:1 is therefore a "higher" gear ratio.

High torque gears (gears with a low ratio) make the life of your motor easier. You should consider the use of high torque gears if:

- you are doing primarily off-road racing on extremely tough track
- your track has plenty of slopes
- you are doing rock crawling

"Gear pitch", on the other hand, can be thought of as the closeness of the gear teeth. A 21DP gear has 21 teeth per inch, while a 30DP gear has 30 teeth per inch (meaning the teeth stay closer together).

Why does it matter? Think about it, assuming you want to change the gear ratio from, let's say, 3:1 to 6:1. If the pitch is the same, the size of the entire gear will have to be doubled in order to accommodate the extra teeth. By keeping the teeth closer, more teeth can be accommodated without increasing the overall gear size (of course, in order to give room for the teeth to stay closer, each teeth will have to be made smaller).

Refer to the picture below, most R/C car gearboxes do not really have room for accommodating significantly larger gears. Therefore changing the gear pitch may be the only viable option in many cases.

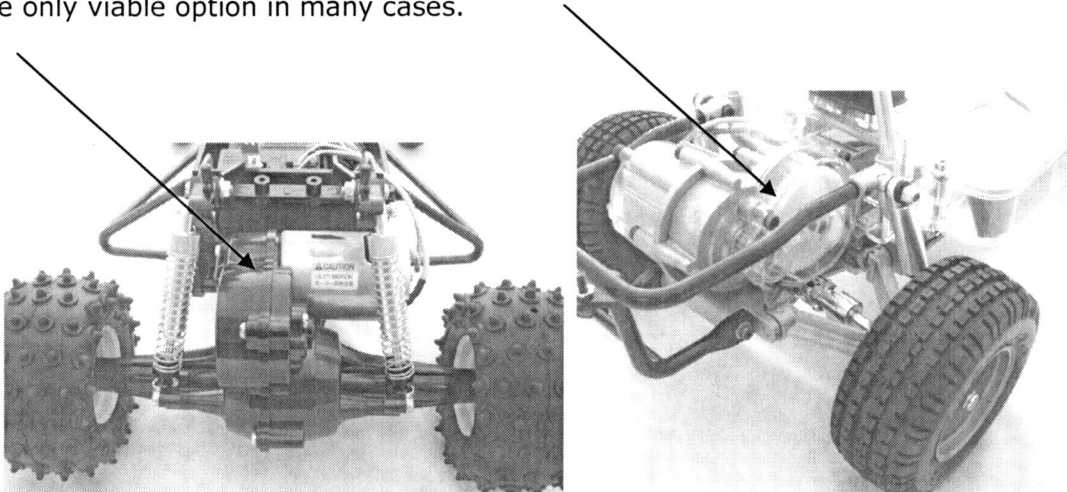

More on Gear Ratios

Gear ratios are usually quoted in the format `of "2.X to 1". This is sometimes represented in writing as 2.X:1. It is simply saying the motor has to rotate 2.X times for the car wheels to complete one full revolution. The more times the motor has to rotate to spin the wheel, the more torque you will get.

4WDs and trucks are tougher on the motor and the batteries, so you'll need to use a relatively lower gear ratio (for the motor to rotate more times). **Generally speaking you can change the gear ratio by changing the pinion and/or the spur gear. HOWEVER, some chassis designs simply would not give enough room for changing to a larger spur gear. Also keep in mind, changing the pinion would require changing the position of the motor, and you simply may not have sufficient room for doing so due to space constraints of the motor compartment.**

The space is severely limited here so you may change to a smaller spur but not a larger one.

Differential gear

A differential serves the following purposes:

- It aims the motor power at the wheels.
- It acts as the final gear reduction in the car.
- It transmits the motor power to the wheels while allowing them to rotate at different speeds.

Why should we allow the wheels to rotate at different speeds? This is because the wheels actually spin at different speeds when turning. In fact, each wheel travels a different distance through a turn, and the inside wheels travel a shorter distance than the outside wheels (thus rotating at a lower speed due to the shorter distance). If the car has no differential gear in place, the wheels will be locked together and spin at the same speed. This will make turning very difficult and awkward - for the car to turn, one tire has to slip.

Each pair of drive wheels needs its own differential gear, therefore a 2WD R/C car has one differential gear while a 4WD has at least two. Some advanced 4WD R/C car has a differential gear between the front and the back wheels due to the fact that front wheels usually travel a slightly different distance than rear wheels. However, with three differential gears in a car, tuning and adjustment can become a quite complicated matter.

There are 2 major types of R/C differential gear. The gear based differential (gear diff) requires almost no tuning and is usually equipped with shaft driven 4WD and 2WD. The ball based differential (ball diff) gives more room for tuning (in fact it requires regular tuning and maintenance) but is said to be more troublesome (again, due to the need for careful tuning). It is often found in belt driven 4WD.

On the DF-03, gearboxes are equipped with ball differentials, and that the propeller shaft is directly linked to the front bevel gear to minimize power loss during transmission.

NOTE: Certain advanced gear diff sets allow you to perform finer adjustment through changing the diff oil inside the differentials. However, most gear diff based electric R/C cars just do not use diff oil at all (they use plain grease on the diff gears instead), meaning further adjustment is not possible.

When you tighten a ball diff to the fullest extent, the differential function is effectively locked.

A gear based differential set (more commonly used for shaft based system):

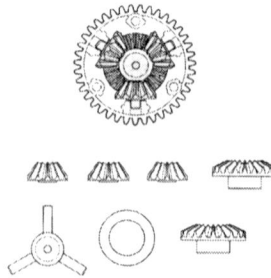

Ball based differentials for a belt driven system:

The front differential

The rear differential

Spool differential

Without the diff fully locked, a crawler will have a hard time crawling through rocks. A spool "differential" actually has no differential action at all. In other words, the wheels are locked to each other. A welded/glued differential has the side gears welded/glued to give the effect of a spool (if you choose to make a spool diff on your own).

If you are using a regular gear diff and you want to glue it tight, simply take it out and have it split opened to where you can see the gears. Apply some epoxy, put the diff back together and let the diff sit over night.

Copyright 2010. **The R.C.PRESS (Hong Kong).**

Why you need to lock the differentials? If you don't, when you start over a rock the differential will send the power to the opposite wheel (the one that is not under load) for no use.

There are also off-the-shelf diff locker parts available but they are usually model specific.

One-Way unit

One-Way is an option commonly found in the high end R/C cars. Its primary function is to allow the front wheels (in the context of R/C, One-Way is primarily for the front) to freely spin even when the motor is not working. Without a One-Way, everything stops when the motor stops. With a One-Way at the front, the front wheels will continue to spin without braking even when the motor is cut off. This feature is desirable because it allows you to maintain traction on the front wheels at the time you brake and steer.

NOTE: With a One-Way installed your car can no longer drive backward (i.e. irreversible movement). The DF-03 MS has a slipper clutch as well as a center one-way system for providing excellent power transmission and high-speed cornering.

Cleaning the gears

If your gearbox has a proper seal, dirt should hardly be able to get in. Regular cleaning is therefore not required. However, the gears will eventually wear out and the lubing grease will need to be replaced accordingly. This is when you should take apart the gearbox and clean the gears one by one. Thoroughly clean the gears by removing all the existing grease and then applying new grease. You may also take this opportunity to flush-clean the ball bearings holding the gear shafts.

The grease that you use must not be corrosive. When you go into your local hardware shop you want to ask for grease that is plastic friendly. Avoid the fluid type grease as they would flow away too easily. You want grease that can stick onto the gear teeth but without damaging them.

Expert tips

- Generally speaking, mixing and matching gears of different brands are NOT recommended because they may not mesh well together. Don't do this unless there is proof that the combination of gears you are interested in will actually work smoothly.

- Don't mesh the motor pinion gear too close with the spur gear or your motor will have a hard time turning. Don't mesh them too loose or gear stripping will result. Fine tuning is necessary to get the motor mounted right. When properly set, you should be able to have the wheels spun forward freely and silently. If the wheels spin with noise, the setting is too loose and will require re-adjustment.

- Make sure the hex screw on the pinion is thoroughly tightened.

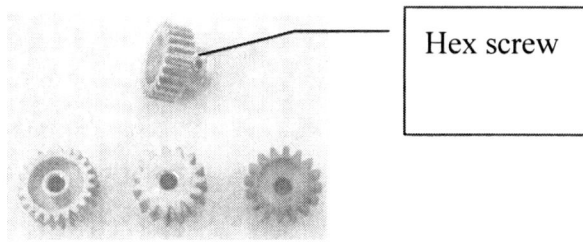

Hex screw

- It takes time for brand new gears to mesh well. Therefore, don't go full speed on a brand new car right from the start. Stick with the stock RS540 motor and run the car at 7.2V for a number of rounds so that the gears can have enough time to kinda "burn in". Switch to a high performance motor only when you find everything running as smooth as expected.

- Ball diff is lighter and smoother given proper tuning. For beginners, however, gear diff is ideal as it is trouble free most of the time.

- You adjust the ball diff primarily through tightening or loosening it. Don't make it any tighter than it needs to be. If there is no slip, you don't need to tighten it any further. In fact, having a ball diff tightened too much (which limits the desired diff action) can be harmful as the balls and rings within it can damage much quickly.

- Slip is the result of a ball diff being too loose. Slipping can lead to power loss, which is for sure not desirable. How do you find out if your ball diff is too loose? Try to rotate the left wheel while holding down the diff case and the right wheel. A properly tuned ball diff should require that you use a certain amount of force for spinning the left wheel. If the left wheel can freely spin without a single bit of resistance, the ball diff is way too loose and you'll need to have it tightened.

- A One-Way may or may not suit your driving habit. It is suggested that you test it out on the track thoroughly prior to having it permanently incorporated into your car.

- If your motor runs but the wheels and the drive shafts do not turn at all, the most likely causes are: 1, the pinion gear is not meshing with the spur gear due to loosed motor mounting screws; 2, the transmission gear is stripped, 3, the pinion is spinning on the motor shaft due to loosed hex screw on the pinion.

- A slipper clutch is a small gear designed to partially disengage (i.e. slip) when the wheels try to drive the motor faster than it would run under its own power. It aims at minimizing the car's braking effect when entering turns. Levant Pro has it included out of the box.

The bearings

Bearings can reduce friction and facilitate smoother spinning by providing smooth metal rollers and a smooth inner and outer metal surface for the rollers to roll against. These rollers bear the load and allow the device to spin smoothly.

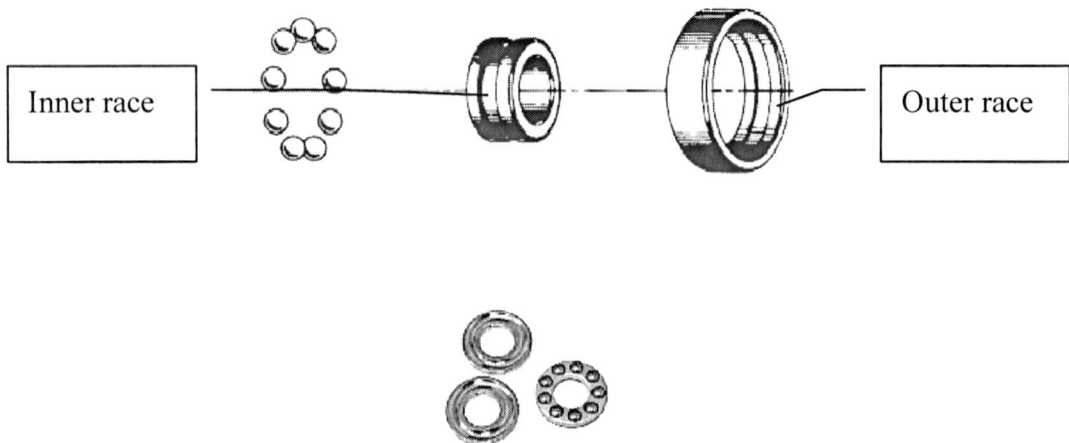

Bearings typically have to deal with two kinds of loading, which are radial and thrust. Depending on where the bearings are being used, they may need to handle all radial loading, all thrust loading or a combination of both. The bearings that support the motor/gear shafts are mostly subject to a radial load.

Ball bearings are the most common type of bearing and are often found in upgraded R/C cars. They use small metal balls as the rollers. When the load is transmitted from the outer race to the balls (and then from the ball to the inner race), the balls only contact the inner and outer race at very small points, meaning there isn't much contact area holding the load. When the bearings are overloaded, the balls can deform or squish. The smaller the balls, the weaker the bearings are.

Cleaning the bearings

The quick and easy way to clean ball bearings is to spray them with silicon lubricant. This will remove most dirt on the outside. When you are done spraying, apply a bit of light oil and let the oil goes into the inner race. This method is easy since you don't even need to take out the bearings. However, cleaning may not be thorough enough (dirt that goes into the inner race can hardly be flushed away this way). Modern teflon sealed bearings offer very good seal so dirt can hardly go into the inner race. Simple flushing on the outside would therefore be sufficient.

Since this "flushing" method is easy to implement, I would recommend that you do this every time after running on a dirty track, to those bearings that are on both ends of the dog bones. On the other hand, those bearings inside the gearbox would not require frequent cleaning (it does not make sense to open up the gearbox every time after a race anyway).

In-depth cleaning requires that you open up the bearings and reach their inner race. Too complicated and too time consuming for beginners and intermediate users...

Special Note: dog bone end float

If you change the stock suspension settings (damper, mounting point ...etc) you may need to check and ensure the dog bones are stilling doing good when spinning. At each end (or on at least one end) of the dog bone there has to be room for "float" and flexibility. However, too much room for float could introduce problem as well. You may find it necessary to add an end pad to each side (or to at least one side) of the dog bone. The end pad does not need to be thick, but it better be made of a flexible material, such as rubber (small rubber O ring will do the job). The goal is to allow for float while filling up the empty room for preventing meaningless "play" of the dog bone.

The pad needs to be fitted into the drive socket. After installation, test spinning the wheel and moving the suspension arm by hand. If movement is difficult, the pad is too thick and you will need to cut it short and retry.

Be ready to increase or decrease the thickness of the pad as you change the suspension setting, in particular the damper (for example, replacing the damper with a longer or shorter one, changing the mounting point ...etc). You can tell the need simply through careful visual inspection and trial spinning of the wheels. If the wheels can't spin smoothly then padding must be reduced accordingly.

Expert tips

- Replace all stock plastic bushings with ball bearings right from the start. Ball bearings allow for smoother rotation of parts and can lead to better energy efficiency.

- Regularly clean bearings that are exposed to dust and dirt (such as those installed on the outer side of the gear case and those attached to the uprights). One quick and easy way to clean bearings is to flush them with silicon oil spray.

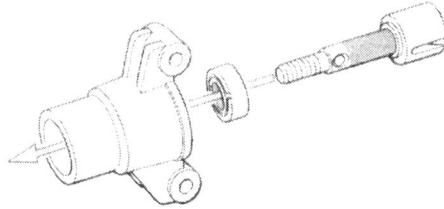

- Ball bearings can break. Therefore, always keep several spare units handy in case replacement is urgently needed.

The motor

Brushed motors

Most off-the-shelf 1:10 R/C car kits from Japan are shipped with the RS540 motor. Certain low power models use the less powerful RS380 motor. Smaller cars like the 1/18 Losi Mini-T use the RS280.

RS540 is a closed endbell motor which is well built and understressed for offering moderate performance balanced against durability:

Newer RTR offerings from the manufacturers in China love to use RC550, which is a slightly more powerful motor than the stock 540.

Copyright 2010. **The R.C.PRESS (Hong Kong)**. All rights reserved.

The diagram and table below show the dimensions and specifications of the Mabuchi RS540RH/SH motors. In fact, you can use these data to establish a baseline for motor performance measurement:

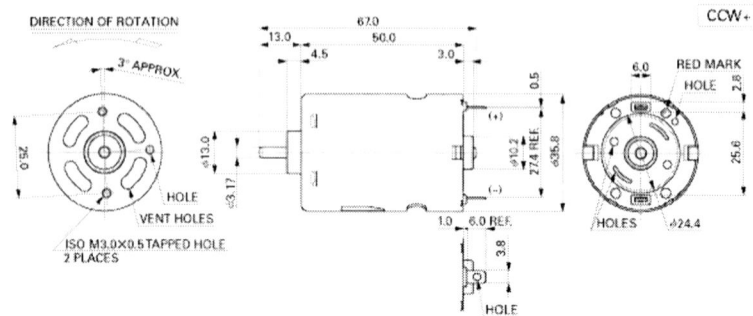

| MODEL | VOLTAGE | | NO LOAD | | AT MAXIMUM EFFICIENCY | | | | |
	OPERATING RANGE	NOMINAL V	SPEED r/min	CURRENT A	SPEED r/min	CURRENT A	TORQUE mN·m	g·cm	OUTPUT W
RS-540RH	4.5 - 9.6	6	11600	0.82	9660	4.09	14.4	147	14.6
RS-540SH	4.5 - 12.0	12	17500	0.95	15080	5.93	31.8	324	50.1

NOTE: Johnson Electrics also produces RS540 compatible motors for the R/C market. The specifications are slightly different though.

Most competition grade motors in the market are designed for substituting the RS540. When looking at their specifications, keep in mind that the "No Load" data is of relatively less importance because in the practical world there is always a load. What you care is how the motor performs under expected load.

Copyright 2010.

Team Checkpoint brushed motors are my personal favorites.

Yokomo's Zero T6 is available with 7T through 12T double-wind armatures.

Motor Turns

Motor turns describe the number of times the wires are wrapped around the armature web of the motor. The less turns a motor has, the faster it will go, and vice versa (so a 13T motor will run way faster than a 20T motor). Motors with very few turns can consume battery power quickly and can also generate massive heat, therefore both your ESC and your battery (and the wires they use) must be strong enough to sustain the heavy load. In fact most modern ESCs are marketed with a rating specifying the "minimum number of turns" that can be sustained.

> NOTE: A lower turn motor such as a 15-T can produce greater RPM but slightly less torque than a higher turn motor. The drawback - running on a low turn motor will require frequent motor maintenance, such as cleaning and brushes replacement. The life span of such motor is generally short as they are way too easy to

Rock crawler requires motor of, say, 60T or even more. Speed is not a concern but torque is. In fact, crawler motor often works with a very high torque (low gear ratio) setting. High lift truck would also require a high torque motor for pushing the heavy weight drive system.

Expert tips

- There are often tradeoffs between torque and rpm. Know what you need: torque or speed or both? In fact, when buying a modified motor, ask the sales person to give recommendation based on the gear ratio and purpose/application of your R/C car (on road VS off road, track condition… etc).

- Powerful motors are often power hungry. You'll need a good battery pack and a real good ESC to support a competition grade motor. If you just upgrade the motor without also upgrade the "infrastructure" of your car, lots of troubles can be expected. If you are less aggressive, going for the general purpose Tamiya sports tuned motor may be a good bet (as it is not that demanding power-wise).

- With the popularity of touring car racing (endurance racing at relatively low speed), there has been a growing demand for higher performance stock motors. The Tamiya Super Stock Motor TZ is one of them. It offers a rpm of 26,500rpm and a torque of 500g-cm (which is reasonably good), both under 7.2V at best efficiency. One good thing about this motor is that it can be easily disassembled for further maintenance and optimization. If you need something real powerful, consider motors from Team Checkpoint (editor's personal favorite).

- Upgrade motors usually come with capacitors for reducing the radio noise created by the motor that can produce radio signal interference and glitching. It is NOT a must have but a recommended option if you are using a very powerful motor. For a stock 540, don't bother to use them.

- You don't need a heatsink on a stock RS540 motor. You do need a heatsink on a tuned high performance motor like the kind shown below. Use a good heatsink if you always race outdoor under high temperature!

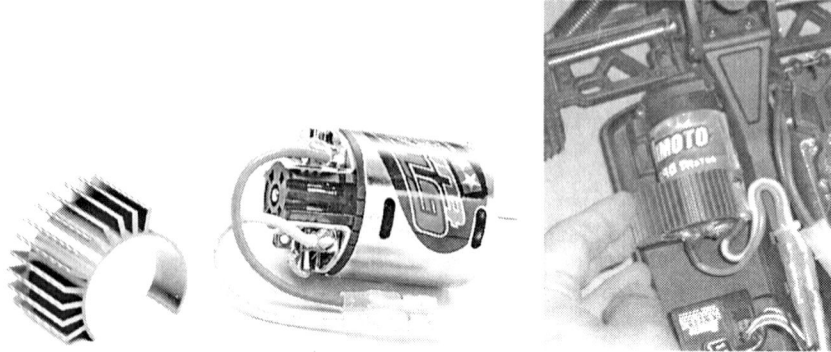

- Installing a motor heatsink is no difficult task. The shape of the heatsink would fit with the shell so all you need to do is to use a little fit of force to "push it in". No need for glue or screw.

- I do not think a fan based heatsink is really necessary. When the car runs, there is enough air flow coming in anyway.

3 Racing has a fan based heatsink available.

Brushless motors

A brushless motor uses an electronically-controlled commutation system instead of the traditional brush based commutation system. In a conventional brushed motor, the brushes make mechanical contact with a set of electrical contacts on the commutator for forming an electrical circuit between the DC electrical source and the armature coil-windings. In a brushless motor, the brush-system/commutator assembly is replaced by an intelligent electronic controller which contains a bank of MOSFET devices to drive high-current DC power and a microcontroller to precisely orchestrate the rapid-changing current-timings.

Brushless motors offer higher reliability, longer lifetime (due to the absence of brush erosion and sparks), and overall reduction of electromagnetic interference. However, they are expensive to manufacture due to the need for high power MOSFET devices in the fabrication of its speed controller and the use of manual labor for winding the stator coils.

Moving from brushed to brushless

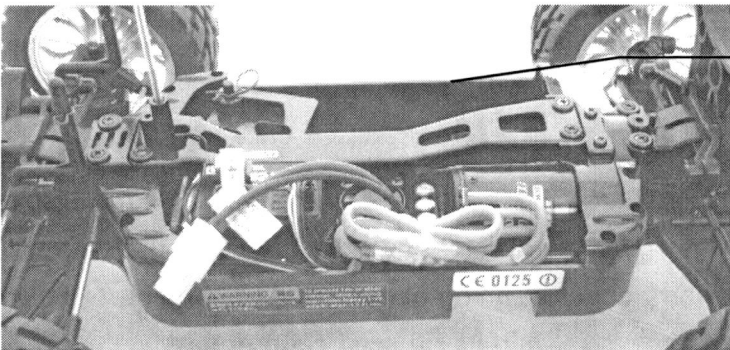

The Levant is brushless out of the box.

All brushless motors for RC cars have 3 wires. The extra wire is used for feedback. Brushless motors require the use of special purpose ESC. These special ESCs has 3 wires for connecting to the brushless motor (traditional brushed motor only needs two):

Copyright 2010.   **The R.C.PRESS (Hong Kong)**. All rights reserved.

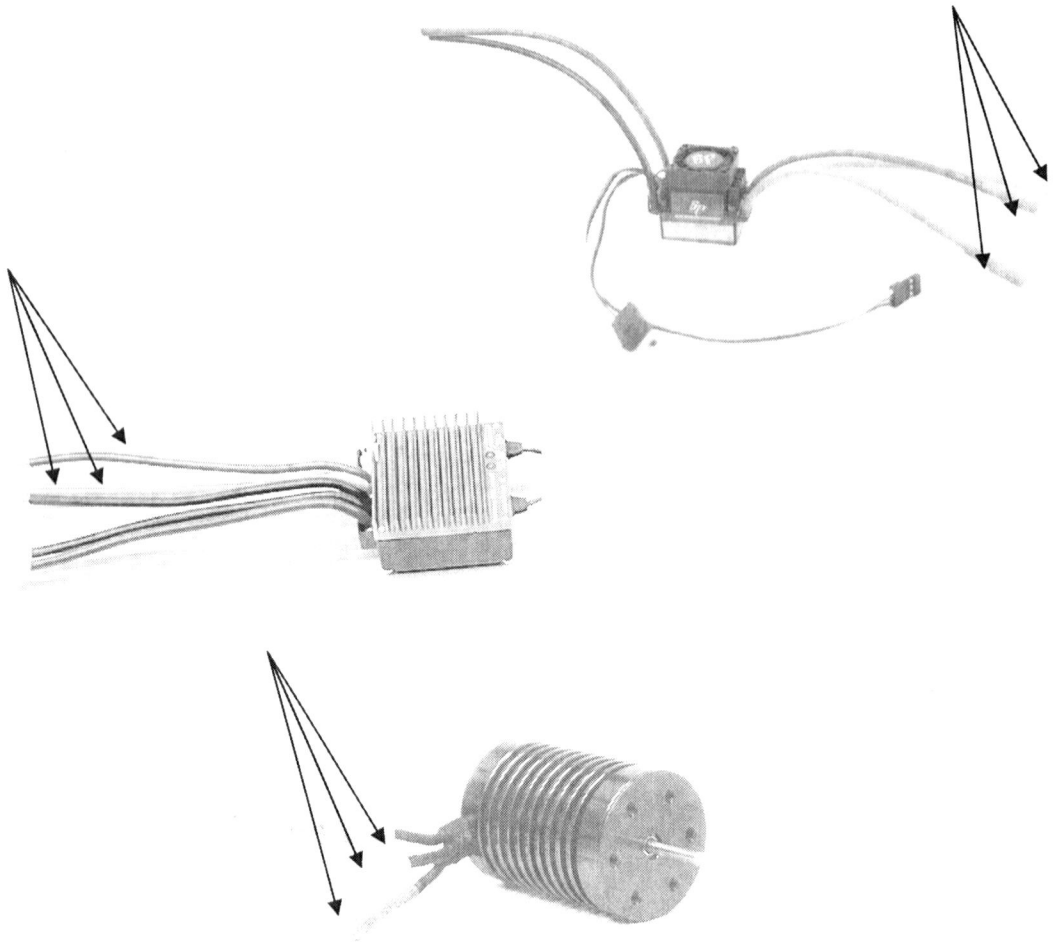

Brushless motor is desirable because there is no friction from the brushes (so it can spin faster and smoother), that there is never the problem of brushes / commutator wear out, and you will never need to worry about cleaning or replacing the brushes. The drawback is cost - it is way more expensive than its brushed counterpart.

If you want to upgrade your RC Car to brushless, you will need to also upgrade the ESC. Now there are combo deals everywhere. Some high end products even come with a programming device for fine tuning the performance parameters:

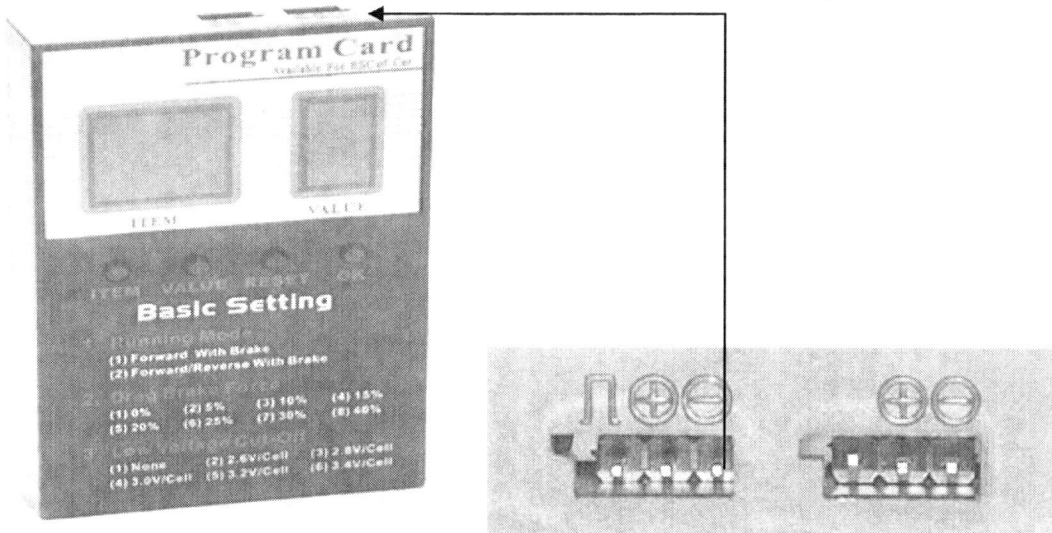

The communication interfaces

Optimizing your brushed motor

The stock RS540 motor has a closed endbell, meaning there is no way you can have it disassembled without breaking the motor. However, proper break-in (of the brushes) and lubing (of the bushings) are still considered beneficial. On the other hand, most tuned / modified motors allow quick and easy disassembly via an open endbell architecture.

Performance optimization

The best way to gain power from your stock motor is to properly break-in the motor brushes and the bronze bushings (most cheaper motors use bronze bushings rather than ball bearings).

You want to first break-in the brush/commutator interface so that the brushes can conform better to the shape of the commutator. To do so is easy – just run the motor on 4 cells (1.2V on each cell) for several minutes until the full brush face is conformed to the commutator.

Before break-in

After break-in

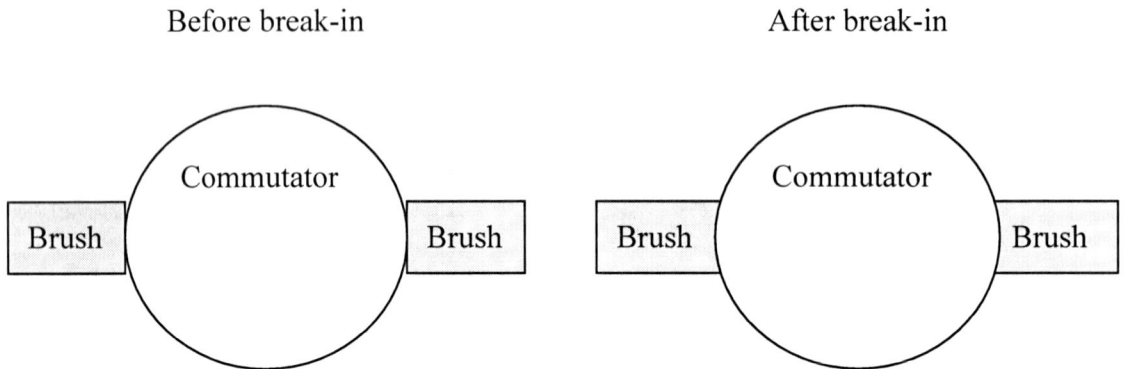

Breaking in bushings is necessary if the motor bushings are too tight. To do a quick check, just spin the motor (by hand) with the brushes removed and fell the resistance to turning (you may want to have some other motors here for comparison purpose).

To perform bushing break-in, just put a little valve of grinding compound into the bushing and spin the motor until you feel a reduction in resistance.

> NOTE: You may find a motor break in stand such as those offered by Team Luna and Kose pretty helpful. With a break in stand you no longer need to hold the motor by hand during break in.

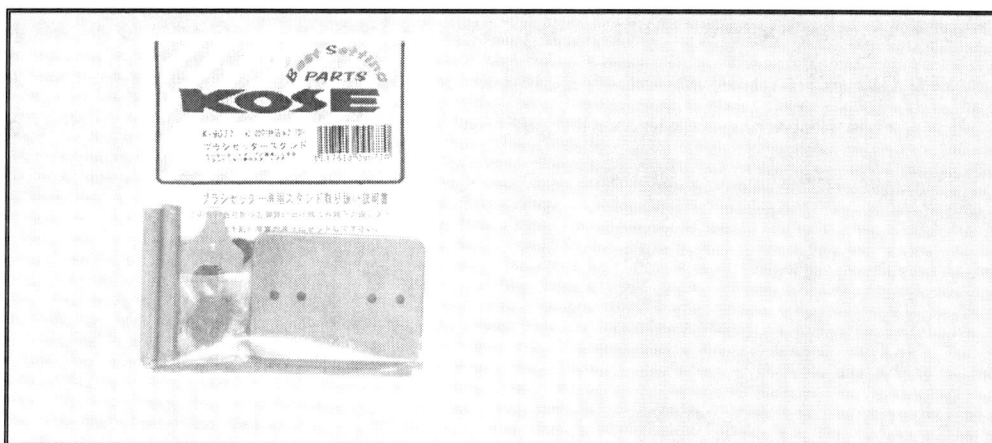

Underwater break in

> Warning: Follow the guidelines here only if you know what "under water"
> !   motor break-in is all about. If you have strong R/C background
>   you should know the tricks. If you don't, ask your R/C buddy to
>   show you how it is done before doing it on your own.

Reason to break-in your motor the "under water" way:

- Brand new motors with new brushes are not perfectly fitted to the commutator. Correct break-in will allow full contact between the brushes and the commutator , which can give less resistance and allow more current to flow through (thus allowing higher torque and rpm).

Why water?

- The water in the container serves two purposes. First it allows the motor to stay very cool as high heat will break down the magnets such that they will not give you the performance that is possible. Second the water will clean the brush particles from the commutator so that you have a clean surface between the commutator and the brush.

Items you will need:

- the Motor
- Old glass jar, cup, bowl or whatever container that is not going to be reused. The deposits from the brushes are toxic and I would not trust it even if well washed. (You can reuse this container for more motors if you wish).
- Motor spray which is available in most R/C hobby stores (break cleaner or any other product with a high flash point will also work)
- A 3v DC power source, such as a couple of batteries (2 "C" or "D" cells in a battery holder works very well)
- Wire or test leads for connecting the battery and the motor
- Light weight oil for lubricating the motor bearings/bushings

Steps to achieve results:

1. Fill the container with water so that when the motor is placed in it the entire motor will be submerged
2. Assemble the 2 batteries so that you can attach wires from the battery to the motor
3. Attach the wires from the battery such that the positive wire from the battery is attached to the negative lead of the motor and vice versa. This will spin the motor backwards from normal operation.
4. Run the motor submerged in the water in this state for a good 5 minutes.
5. Take the motor out and dump the water.
6. Refill the container with fresh water and connect the batteries such that the positive contact of the battery is connected to the positive lead of the motor. Do

the same with the negative lead from the battery so that it is connected to the negative contact of the motor.

7. Run the motor submerged in the water for about 10 minutes.

8. Dumb the water and rinse out the container to be used again for further motor break-in (NEVER use this container for food or other substances which you eat or drink).

9. Take the motor spray and totally clean out the inside and outside of the motor such that there is no water left. Since the motor cleaner has a high flash point, it will clean away any water left inside the motor.

10. Now take the light weight oil or lube of your choices, apply a small amount to both bearings/bushings.

11. Spin the motor in your hand for a minute or two to make sure everything is running smoothly and to work in the lubrication into the bushings/bearings.

Motor bushing break in

To perform motor bushing break-in, just put a little **drop** of grinding compound into the bushing and spin the motor until you feel a reduction in resistance. Don't forget to spray out the dust particles from the brushes with motor spray (or other high flashpoint spray solvent). If these particles are not cleaned off they will further wear away the brushes and the commutator.

Adding torque to the motor

You can add more torque by adding a "Torque Sleeve", which can be easily ordered from your local R/C hobby stores. It works by effectively shielding the magnetic field of the magnets in such a way that more of the field stays inside the motor can rather then radiates outward away. Below shows a torque sleeve manufactured by Trinity:

Many brushed packages come stock with a torque sleeve:

Copyright 2010. **The R.C.PRESS (Hong Kong)**. All rights reserved.

Torque sleeve

Another quick and cheap way to increase the torque of your stock motor is to replace the stock motor springs (the springs that hold the brushes up against the commutator). Harder springs produce more tension and more amp draw, thus resulting in more torque but less RPM. On the other hand, if you want less torque but more RPM, a pair of softer spring is the way to go. Take the motor to a local electronic shop and find springs that work for you. Do keep in mind that very hard springs are no good as they impose too much pressure on the commutator. Very soft springs are no good as well, since they allow the brushes to bounce on the commutator. Some trial and error may be necessary to get the fine-tuning done properly.

Motor maintenance

Keep in mind that excessive heat will weaken the magnets of the motor. It is not uncommon for racing motors to get overheated and stop working. Most of the time an improperly configured gearbox is the problem (gears too tight or something get stuck inside). If the outdoor weather is very hot (to an extent where yourself cannot afford staying inside a vehicle without A/C), give the car a break after every 5~10 minutes. Remove the motor cover (if any) and give the motor enough time to cool down.

> Warning: Heat is harmful to all DC motors (all R/C motors are DC based) because of 2 reasons. Firstly, heat can permanently degrade the magnetic field produced by the magnets. Neodymium magnets will de-magnetize entirely at around 200 degrees Centigrade.

> Extreme temperature changes, not necessarily high heat but just the change of temperature, will also degrade the magnetic field. Secondly, heat in the windings can cause the copper wire to have a higher resistance and deliver less rpm and torque out of the motor. This in itself is not harmful to the motor windings, just that you will see less performance.

With proper maintenance, your stock motor should be able to run at maximum power and speed. Below are what should be proactively done:

- Every time after a one-day racing event, use a high quality motor spray to clean out the dirt and the carbon deposits.
- Replace the motor brushes regularly (check the brushes. If the tips that touch the commutator have turned purple, replace them immediately).
- When you change the brushes, consider changing the brush springs too as they might have lost their tension already.

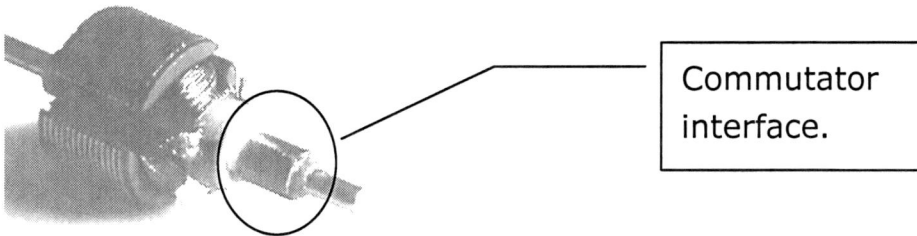

Commutator interface.

Using 11.1V to power your vehicle

Using a 11.1V Lipo on your vehicle can for sure deliver excellent horse power. HOWEVER, it can fail your ESC in minutes. If you are using high power battery like that you want to keep in mind the following points:

- A fan blowing at the solder posts of the ESC can help a lot. 5V or 12V DC fans used by computer coolers are fine (you don't need 12V to push a 12V DC fan... the fan will just spin slower with a lower input voltage). If you use a 5V fan you want to get power from the ESC-to- receiver connection (through the red and black wires) or from a separate source that does not go higher than 5V.

- A large heatsink always makes a huge difference in the amount of constant power the ESC can handle. If you are to custom make one, use aluminum as the material, which is cheap and easy to cut.

- The BEC gets weak as the supply the voltage goes up. Under a 11.1V configuration a separate BEC may be necessary.

So what is a BEC? It is the abbreviation for Battery Eliminator Circuit, a built-in voltage regulator that supplies a constant voltage to the receiver and the servo. The receiver does not need an external battery pack, but one can be used if needed.

Some ESCs provide a switching BEC that can handle higher input voltage. Tekin's Rx8 is an example.

It is also known that many use 11.1V lipos for their brushed configurations on the Tekin FxR without any problem.

Sensored BL VS Sensorless BL

What is the difference between a sensored BL system and a sensorless system? Sensored is all about the connection between a brushless ESC and motor which allows the ESC to be informed of the precise position of the rotor in relation to the windings in the motor. This allows for proper low speed drivability and braking. A sensorless system is less costly to manufacture but low speed drivability is relatively poor. In terms of top speed there is in theory no difference though.

The Tekin R1s are sensorless while the R5s are sensored:

The suspension system

Overview

The job of the R/C car suspension is to maximize the friction between the tires and the road surface and to provide steering stability with good handling, which in turn allows for a smoothly controlled ride.

Generally speaking, the four wheels of a R/C car work together in two independent systems, which are the two wheels connected by the front axle and the two wheels connected by the rear axle. It is technically possible for a car to have totally different type of suspension on the front and back. Take the classic Tamiya Hornet as an example, it has a rigid axle that binds the rear wheels (a dependent system), but the front wheels are permitted to move independently (an independent system).

NOTE: Both the Hornet and the Grasshopper have a bathtub style chassis for holding a simple rigid axle in the rear. Due to this simple architecture energy transmission across the drive system is extremely efficient. That was why Hornet was real fast even with just a 540 stock motor.

A dependent system has a simpler structure but is less effective for shock absorption and traction control. In fact, dependent rear suspensions have not been used in mainstream R/C cars for years.

Most modern R/C cars allow each of their wheels to move independently.

You should always go for R/C cars that come out-of-the-box with independent systems front and back. On a regular R/C car, changing from a dependent system to an independent one is almost impossible.

For crawler, however, dependent system is used both front and back. In order to traverse difficult rock formations, the crawler needs to twist. This requires the use of dependent system (for twisting to take place).

Components of a suspension system

The primary components of a suspension system are dampers and stabilizer bars (anti-sway bars).

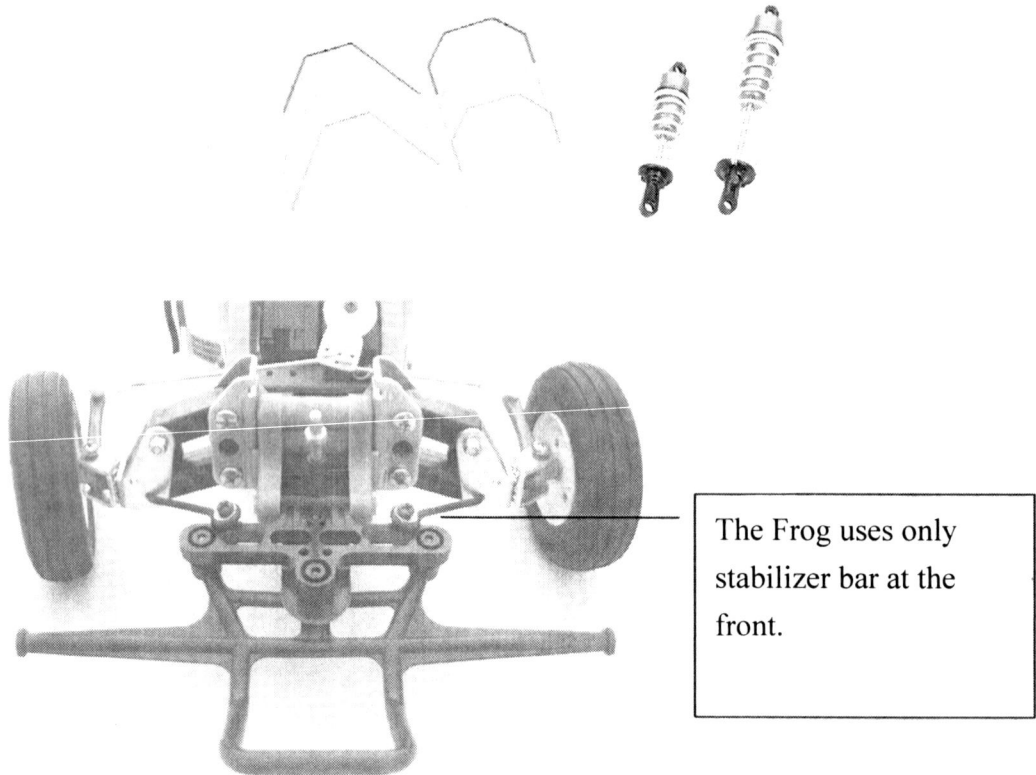

The Frog uses only stabilizer bar at the front.

In the context of vehicle dynamics, a ride refers to a car's ability to smooth out a bumpy road. Handling, on the other hand, refers to a car's ability to safely accelerate, brake and corner. A good suspension system is expected to achieve the following:

Road Isolation, which refers to the car's ability to absorb or isolate road shock, thus allowing the car body to ride undisturbed while traveling over rough roads. To achieve this, the suspension system must be capable of absorbing energy from road bumps and dissipating it without causing undue oscillation in the car.

Copyright 2010. **The R.C.PRESS (Hong Kong)**. All rights reserved.

Road Holding, which refers to the degree to which the car maintains contact with the road surface in various types of directional changes and in a straight line. To achieve this, the suspension system must try to keep the tires in contact with the ground as much as possible through minimizing the transfer of vehicle weight from side to side and front to back (as this transfer of weight can reduce the tire's grip on the road).

Cornering - the ability of the car to travel a curved path without body rolling (roll tends to put more weight on the outside tires and less weight on the inside tires, which can reduce traction and mess up steering totally). To achieve this, the suspension system must be able to transfer the weight of the car during cornering from the high side of the vehicle to the low side.

The dampers

A damper is basically a tube like device placed between the chassis of the car and the wheels. The upper mount of the damper connects to the frame, while the lower mount connects to the axle near the wheel. The 2 major types of R/C damper are the oil filled shock absorber and the pure spring based damper.

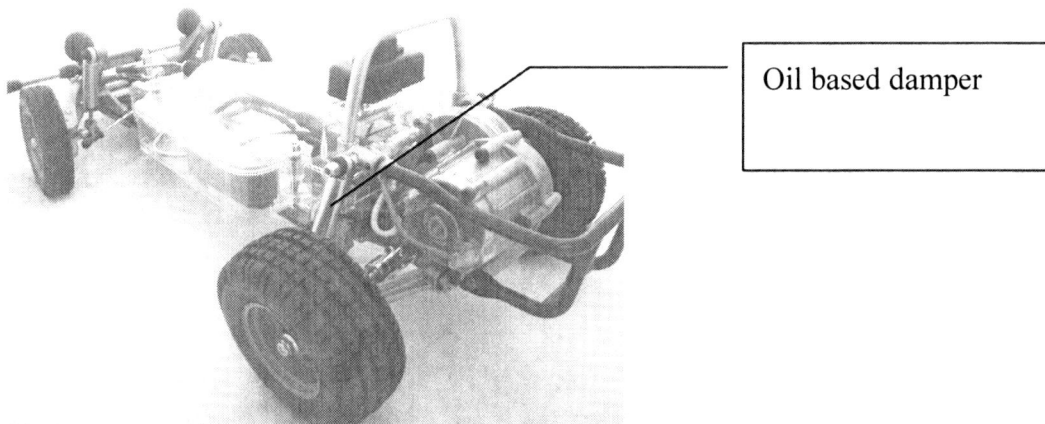

Oil based damper

Copyright 2010. **The R.C.PRESS (Hong Kong)**. All rights reserved.

Oil based damper

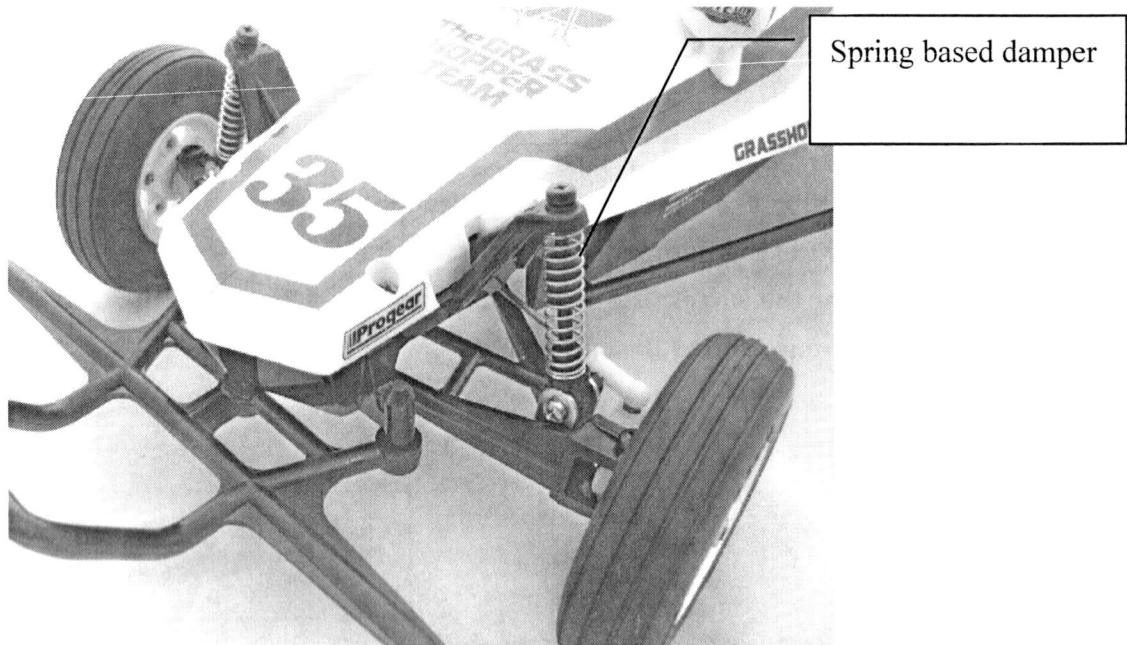

Spring based damper

With a pure spring based damper, you fine tune suspension through adjusting the stiffness of the spring. An oil filled shock absorber, on the other hand, allows for smoother response to differing road conditions through slowing down (smoothing out) the movement of the spring (the "smoothing out" effect can be fine tuned through using different grades of damper oil, which is commonly described by thickness but measured in weight – 10wt is the thinnest while 100wt is the thickest). Generally speaking, an oil filled shock absorber performs better in terms of road holding and road isolation.

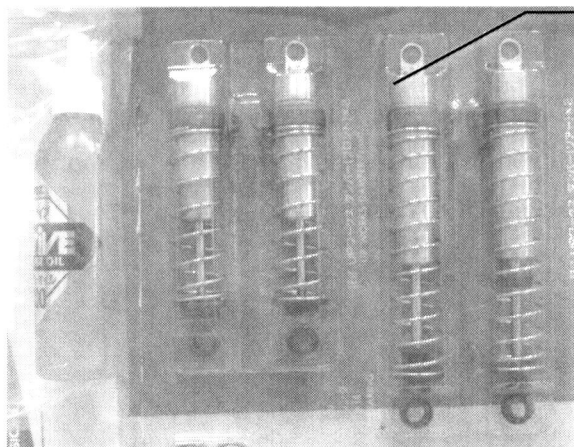

Advanced damper set with damper oil

NOTE: An oil shock absorber has a relatively complicated internal structure with parts like piston, piston rod, valve, pressure tube and cylinder involved, thus leading to a much higher manufacturing cost (the lower cost Tamiya R/C cars use pure spring based dampers due to this reason).

When the wheel of your R/C car encounters a bump in the road which causes the damper spring to coil and uncoil, the energy of the spring is transferred to the shock absorber through the upper mount, down through the piston rod and into the piston inside the oil shock absorber. Orifices perforate the piston and allow oil fluid to leak through as the piston moves up and down in the pressure tube. Because the orifices are relatively tiny, only a small amount of oil fluid can pass through. This slows down the piston, which in turn slows down the spring. Regular cleaning is necessary or the damper will get oily. If there is oil leak you may need to replace the small rubber O ring inside it.

Double wishbone VS single wishbone

A double wishbone suspension has both an upper arm and a lower arm on each side. Such an architecture is more solid and sturdy when comparing to a single wishbone design commonly used by the earlier generations of R/C buggy:

The DF-03 has 4-wheel double wishbone suspension with long suspension arms.

A double wishbone design:

A single wishbone swing arm design (there is no upper arm. Instead, a simple rod is used to hold the upright):

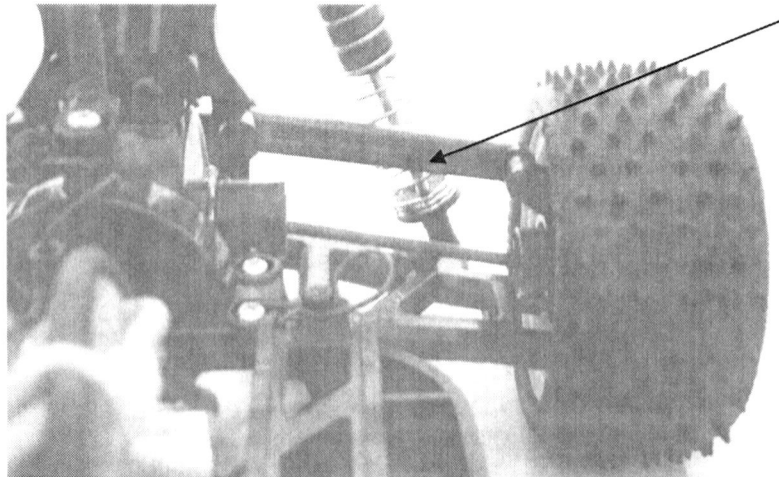

Fine tuning the shock absorption function

Technically speaking, a sprung mass refers to the mass of the car supported on the suspension springs. The stiffness of the springs and the shock absorption function affects how this sprung mass responds while the car is running on the road. Loosely sprung cars (those with "softer" suspension) can swallow bumps for smoother riding but is prone to dive and squat during braking and acceleration and may easily roll during cornering. Tightly sprung cars (those with "harder" suspension) are less capable of swallowing bumps on bumpy roads but can minimize body motion better, which is more suitable for an aggressive driving style.

If a pure spring based damper is in use, you may adjust its stiffness through spring replacement or through spring compression with ring type spacers. If an oil filled shock absorber is in use, you may also want to change to a different damper oil or to use travel limiters (which are spacers that come shipped with most absorber kits) to limit the movement of the absorber. Consult your local hobby store for parts availability.

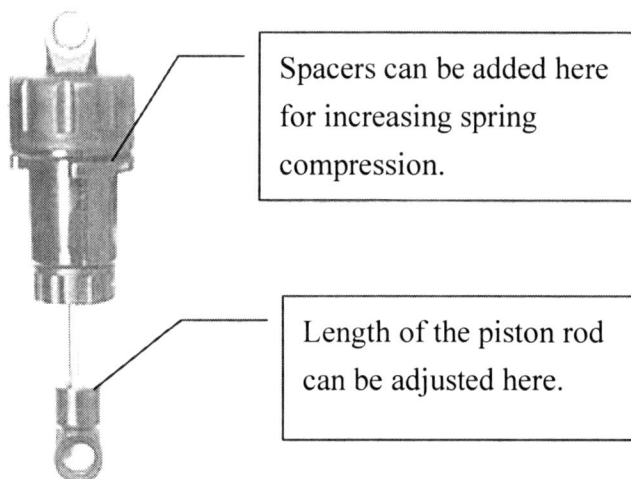

Spacers can be added here for increasing spring compression.

Length of the piston rod can be adjusted here.

Certain Tamiya touring car models like the one shown in the picture below uses a single horizontal damper for suspension on both end, which may be sufficient only on relatively flat race tracks:

For buggy/truck racing, it is always preferable to use independent damper for each wheel.

The angle of the damper can also make a difference. Some R/C car models allow you to fine tune the mounting angle of the dampers independently (for example, the Tamiya TT01R chassis kit's lower arm features a choice of 2 locations for damper attachment), while others may require that you drill holes on your own to change the mounting positions:

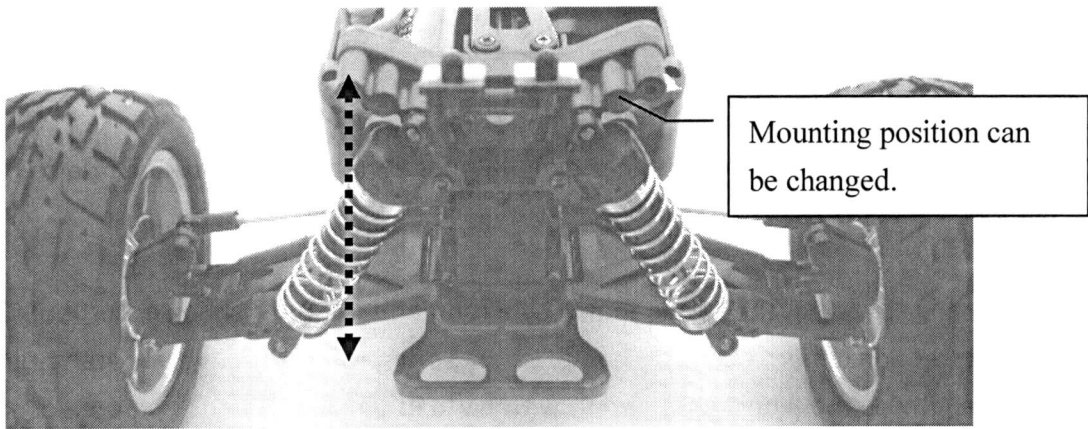

Mounting position can be changed.

It is believed that with almost vertically mounted dampers you can have much better response when turning (at the expense of traction), and vice versa. Again, differing road conditions can produce different results in relation to how the dampers are mounted. You may want to install a special damper mount to give room for further adjustment and tuning. 3Racing offers several different mount towers for different Tamiya chassis platforms:

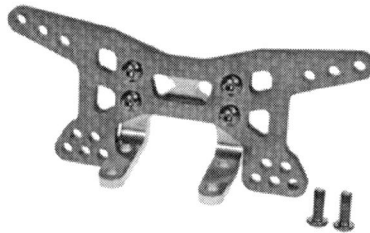

For a 1:10 crawler, the typical damper length is between 80 - 130mm. Generally, you should want your damper spring to be as soft as possible to enable maximum axle articulation. Axle articulation refers to the car's ability to flex its suspension.

Making your own damper springs

If you are going to tailor make your own damper springs, the following information on compression spring will be useful for you.

Technically speaking, the spring in use by a regular damper or shock absorber is known as compression spring. On every compression spring there are several parameters. The "free length" is the length of a spring with no load applied. "Length before set" can be thought of as near the maximum compressed length of a spring (let's refer to this as the compressed length). The number of coils also plays a role in determining the strength of a spring. Generally speaking, the more coils a spring has the more powerful it is. However, the material that is used for producing the spring and the corresponding heat treatment process are even more critical. Put it this way, a soft spring with 10 coils may well be "weaker" than a hard spring with 5 coils. Also, not all coils are active (this apply especially to variable-pitch springs). And, installation-wise, a spring that is too long (a high free-length parameter) is much harder to fit into the damper. Keep in mind that:

● when installed, both ends of the spring (without compression) should not have any gap left. If there are gaps on either side, you may need to use spacers. HOWEVER, the need for many spacers may simply indicate that the spring you selected is too short.
● note the diameter of the spring wire. If the spring wire has a large diameter, the compressed length may be too long (which can limit the movement of the damper piston).
● note the number of coils. A short spring with MANY coils can be very lengthy even at full compression.
● always have your custom made springs properly ground.

Grinding is the process of grounding the ends of a spring.

Generally speaking, when a spring is ground on both ends, it can spin more freely during compression.

NOTE:	Do it yourself grounding: It is very easy. Just take the spring to the kitchen, turn on the fire, have the end of the spring heated, and then ground it by force.

The use of stabilizer bar

Stabilizer bar (anti-roll / sway bar) can keep the car's body flat by moving force from one side of the body to another, thus reducing the chance of body rolling in a sharp turn (and also reducing the traction you have during a turn). However, most stock R/C cars do not come equipped with one. In fact, stabilizer bar packages are usually available as optional upgrade parts. Check with your local hobby stores for availability.

Soft springs with stiffer stabilizer bars and stiff springs with softer bars can accomplish the same thing - that is, to provide a stable suspension while reducing excessive weight transfer during a break (there is a tendency for weight to get shifted as you break).

Expert tips

- A stiffer suspension setup can effectively limit the movement in the suspension, which in turn can allow for a more stable straight-line ride. A "softer back" (a softer rear suspension) and a "stiffer front" (a harder front suspension) together can allow for more aggressive steering (it allows you to turn more in a more responsive manner). On the other hand, a "stiffer front" together with a "stiffer back" tend to give stable straight-line ride at the expense of traction.

- The stiffness of your suspension system can be fine tuned for achieving a proper ride height. It is perfectly fine if you intentionally make either the front or the rear stiffer. In fact, a stiffer rear suspension can lead to a higher ride height at the back, which allows for more steering. HOWEVER, it is NEVER okay for one side of the car to go stiffer than the other side (we are talking about left/right stiffness here). Left/right balance must ALWAYS be maintained.

- Traction is usually good on carpet surface, meaning you can go with a stiffer suspension setup. On tracks that are extremely bumpy:

  - use of thinner damper oil and softer springs can allow the shock absorbers to take more abuse.

  - use of thicker damper oil can allow for better control after "landing" from jumps.

  - use of harder damper springs can make your car extremely jumpy and bumpy (a hard spring is harder to compress but tends to rebound much "harder" too).

- Braking will always cause weight transfer towards the front of the car and reduce rear traction significantly.

- Stiffer suspension can lead to faster tire wear. Exactly how much faster can vary depending on the road condition as well as on the type of tires you use.

- One easy way to determine if your existing dampers are too soft for the track is to check the bottom of the chassis after each ride and see if new scratches are popping out all over the place. To determine if proper left/right balance is maintained, see if these scratches are evenly distributed.

- Through changing the damper oil you may easily control the dampening characteristic of the piston within a shock absorber. Although it is technically possible to switch to piston with different size holes or different outer diameter, you don't really have to go this far at this level. If oil is leaking out of your shock absorber, most of the time it is just that you have over-filled them. The rubber O ring inside the absorber seldom breaks by itself.

- It is not unusual for the piston rod of the shock absorber to get oily after a ride or two. Make sure you keep it clean all the time as dirt often likes to stick to the oily surface.

- Check the oil level of your shock absorber regularly. Insufficient oil inside the absorber can lead to slower rebound. Refill if necessary. AND make sure you put in the same amount of oil for the left and right absorbers! Also, immediately check the suspension system after each crash. The piston rod may be bent during a crash.

- You need to install stabilizer bars only if your car always rolls during sharp turns. If your car does not roll due to good driving habit or that the track you race on does not require frequent sharp turns, then you don't really have to install stabilizer bars.

- On a RWD truck you generally do not need stabilizer bar at the front.

- When installing a stabilizer bar, make sure you have it properly aligned without leaning towards either side. Also make sure that the rods for connecting the bar to the suspension arms are properly adjusted. An improperly aligned stabilizer bar can lead to inconsistent left/right steering.

- Since the use of stabilizer bars can affect the way your car steers, you should try out bars of different stiffness until one which suits your driving style is found.

- A front-only stabilizer bar configuration, a rear-only stabilizer bar configuration and a both-ends stabilizer bar configuration all produce different steering effects. You should try them out one by one to find an ideal configuration that suits you most.

- Check the shock mount/shock tower. Some R/C cars come with ABS plastic shock mounts which like to flex all the time. You'll need to either reinforce the flexing mounts manually (which can be tricky) or use off-the-shelf/tailor-made replacement mounts produced with stronger material such as aluminum alloy.

The wheels and the tires

Camber, caster and toe angle

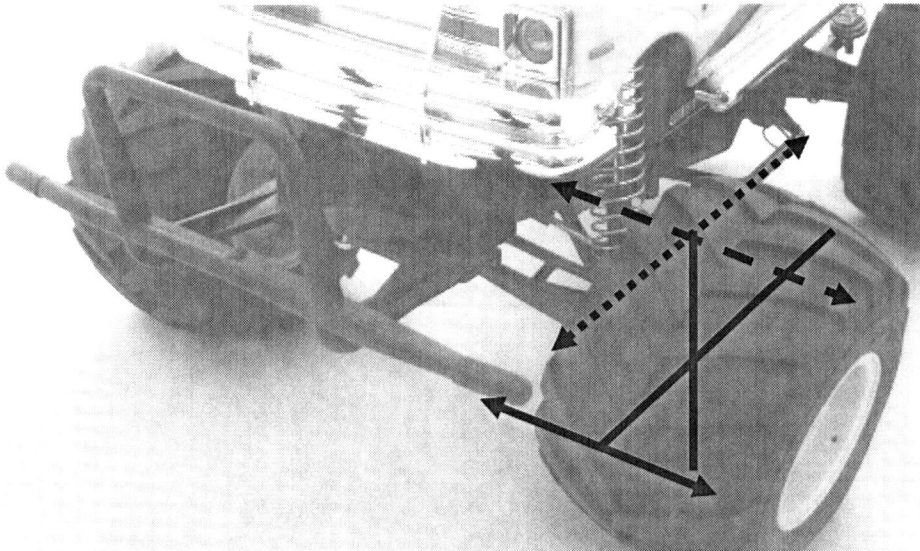

Camber: ◀ ─ ─ ─ ─ ─▶

Caster: ◀••••••••••••▶

Toe angle: ◀───────────▶

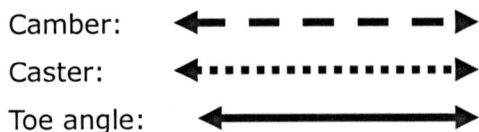

Talking about wheels, your primary concerns are the camber, the caster and the toe angle. Camber refers to the vertical angle of the tire in relation to the ground. Camber adjustment aims at controlling how much tire and traction is grabbing the track surface at any moment. When cornering or making a sharp turn, the natural tendency is that your car's outside tires will lean towards the outside of the turn. In order to keep as much tire on the track surface as possible, you may opt for a negative camber which involves making the angle of the tires leaning inwards towards the chassis. Positive camber, on the other hand, is rarely used. There is a pitfall - too much of a negative camber may lead to the loss of traction on straight line. Therefore, you must carefully adjust the camber based on the unique track condition you are dealing with. Ask yourself one simple question: are you going to have to do lots of cornering on this particular track?

Many R/C cars allow camber adjustment through adjusting the upper arms (if the upper arms are adjustable – certain lower end 1:10 cars do not have adjustable upper arms).

Caster is all about the angle at which the front steering pivots. It refers to the forward (negative) or backwards (positive) tilt of the steering axis. Positive camber will provide for less grip. The diagram below should sufficiently explain:

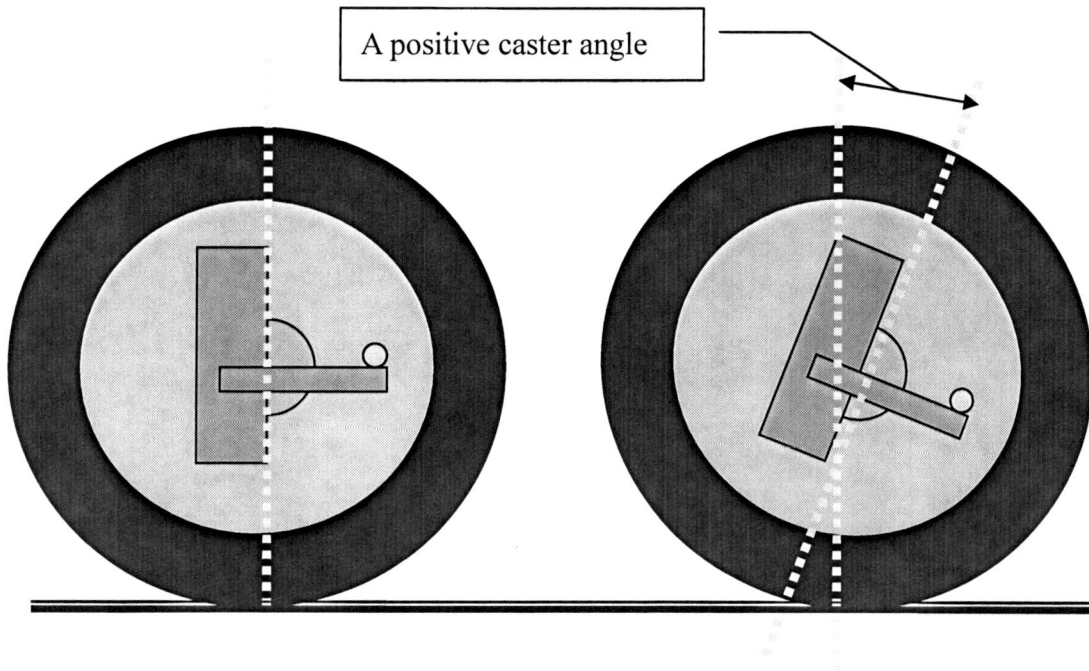

A positive caster angle

Caster has the progressive effect of leaning the front wheels into the direction of the corner. Therefore, the more the caster angle is laid back, the greater the effective camber change when you turn the wheels. **However, caster adjustment is NOT allowed on most hobby grade electric R/C cars.**

Toe Angle refers to the angle with which, looking down on your car from the top, the wheels angle in toward the front. This is one of the most commonly adjusted settings on a R/C car.

Toe-In means the wheels are pointing inwards, while Toe-Out means the wheels are pointing outward. Adjustment on the front is done primarily through adjusting the length of the steering rods (if they are adjustable).

Toe-in can lead to under-steer. Toe-out can provide more aggressive steering response. Zero toe can reduce tire wear.

Certain R/C car models also allow rear toe angle tuning through rod adjustment at the back:

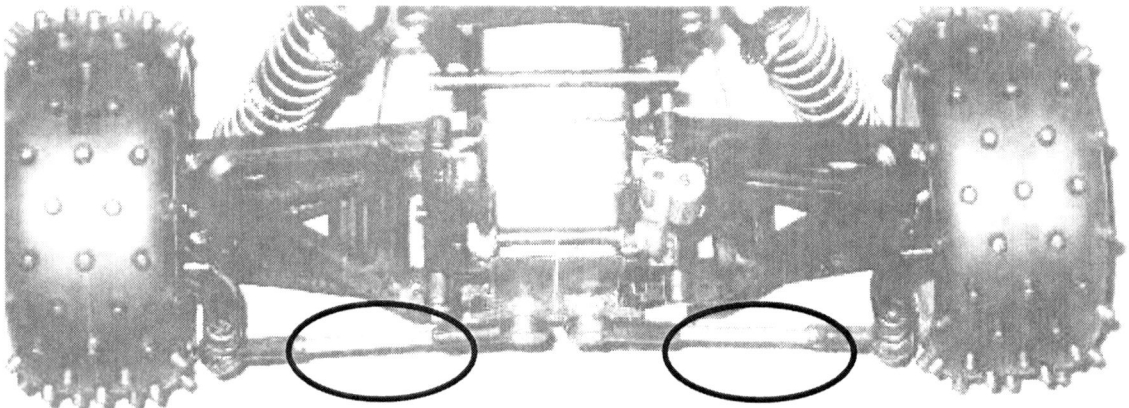

The good thing about adjustable linkage rod is that it is flexible and tunable. The bad thing is that overtime it will bend (at the joints). One piece linkage rod is not adjustable but is much stronger.

Most Tamiya offroad EP models do NOT allow rear toe angle adjustment by default.

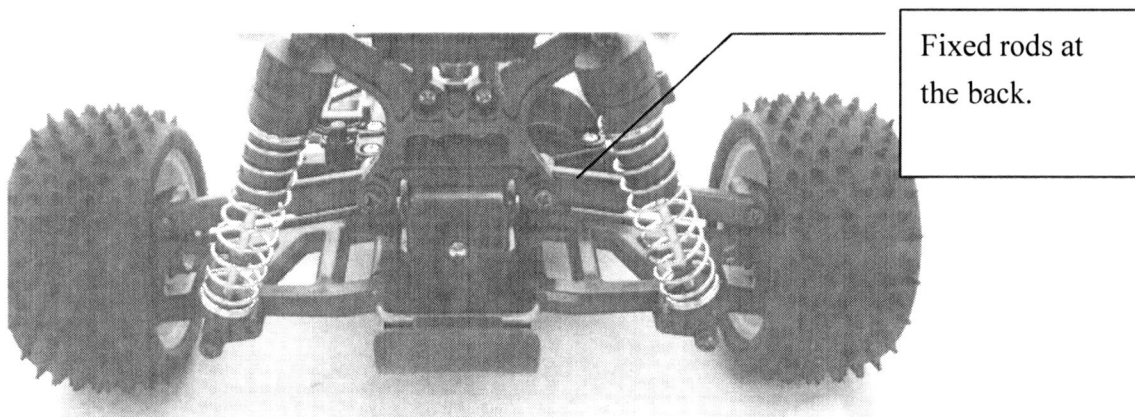

Fixed rods at the back.

Types of tire and their use

Performance tires are designed for use at higher speeds. They often have a softer rubber compound for improved traction (which is necessary for high speed cornering). The ultimate variant of these tires has no tread pattern on the surface and is called slick tire.

Slick tires have no treads on them and are ideal for touring cars which run on flat, dry and clean tracks where reasonable traction can be expected. Tread tires have treads for channeling down any water or dust encountered on the track and can provide maximum grip out of the ground. There are hundreds of tread patterns available, although the actual pattern itself is mostly a mix of functionality and aesthetics. Pin tires are solely for off road racing where the tracks are bumpy and dusty (or wet) or that traction is totally absent. Because the pins can wear out real quick on flat surface, you don't want to use them for on-road racing.

Mostly used by R/C trucks, mud terrain tires have large, chunky tread patterns designed to bite into muddy surfaces and provide grip as well as to allow mud to clear more quickly from between the lugs. These tires tend to be wider than other tires for spreading the weight of the truck over a greater contact patch and may not well suit to on-road use.

Some tread designs are unidirectional, meaning the tire has a desired rotation direction for enhancing straight-line acceleration through reducing rolling resistance. It is therefore important for you not to put a 'clockwise' tire on the left hand side or a 'counter-clockwise' tire on the right side (check the tire's manual – it should tell you the correct installation direction). Symmetrical tire, on the other hand, has a tread pattern consistent across the tire surface (meaning both halves of the treadface are of the same design) and is therefore not direction sensitive installation-wise.

Typical offroad tire tread patterns.

Crawler uses over-sized, low-pressure, knobby, mud-terrain tires.

Crawler needs traction badly so a soft foam setup on the tires is usually required (especially in cold weather).

Tire performance factors

For improved overall performance and extended tire tread life, it is imperative that the tires be in proper alignment with the R/C car. Poor or improper alignment tends to occur when the suspension and steering systems are out of adjustment.

The size and dimension of a tire can also make a difference. If you use larger tires than specified without changing the gear ratio of the car, running performance (speed, torque and energy consumption) as well as steering capability can be seriously affected, depending on how larger the new tires are.

All tires work best when warmed. This explains why your tires always give less grip on the first several rounds. There are manufacturers which produce "tire warmer" for the R/C community (like the one shown in the picture below). You may also warm your tire manually via the use of hair dryer (do this only when it is very cold).

The Alex Racing Black Heater 4 Tire Warmer Kit:

When installing and changing your tires, make sure you maintain proper balance on them. Radial Run-out refers to an "out-of-round" situation where vibrations are produced as the wheel spindle moves up and down. A lateral Run-out, on the other hand, refers to a side-to-side or wobbling movement of the tire and wheel. Tire imbalance is usually the result of improper tire/wheel installation, bent wheel frame or deformed tire surface. Imbalance is easily detectable through visual inspection.

> NOTE: The technical definition of balance is the uniform distribution of mass about an axis of rotation, where the center of gravity is in the same location as the center of rotation. A balanced tire is one where mass of the mounted tire is uniformly distributed around its center of rotation. Imbalanced tires can lead to annoying vibration - this symptom tends to increase in magnitude with greater speeds.

Expert tips

- Car handling is a subject which involves complex interaction of MANY different factors (camber, caster, toe angle, shock absorbers, spring rates, stabilizer bar and the like). Therefore, no single adjustment can turbo charge your car. However, for ease of troubleshooting, it is suggested that you tune these elements one by one so you can tell what makes a difference and what does not.

- One easy way to tell if something is wrong is by checking the tires. If tire wear is extremely uneven, something must have been incorrectly set. **Misalignment of the wheels due to improper settings can cause either the inside or the outside to wear unevenly, or to have a rough, slightly torn appearance. Poor alignment often results in excessive and/or uneven tire wear. Improper alignment can reduce a tire's usable life by as much as 70~80%.**

- You control the over/under steer effect of your car through Toe Angle adjustment:
    - More front Toe-In can lead to under-steering but can produce better straight line stability.
    - More front Toe-Out can give extremely aggressive steering response for corning at the expense of straight line stability.
    - Rear Toe-In improves straight line stability at the expense of low speed cornering.
    - Rear Toe-Out is rarely used.

- With the front wheels set with toe out, the car can turn quickly and can get into a corner as fast as possible. With the rear wheels set with toe out, the car can get into a slide easier.

- When attaching the tire to the wheel, don't forget to apply instant cement around the tire (so the tire can stick firmly to the wheel frame).

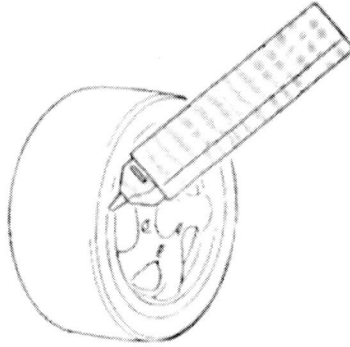

- Remember to have a balanced left/right setting. The setting on the left should NEVER differ from the setting on the right. This holds true even for a short course truck.

- Aluminum wheels look cool (due to the metallic feel) and are sturdier. However, they are heavy (thus increasing the workload of the motor) and can increase the likelihood of bending axles/breaking arms when raced hard. Therefore, stick with the plastic wheels whenever possible.

R/C electronics

The remote controller set

The most popular controller set for R/C cars is the wheel-type 2-channel system. With a 2 channel system you can control steering with one channel (channel ONE) and throttling with another (channel TWO). In fact, there are two possible configurations using a 2-channel system:

- Configuration one: a wheel type transceiver + ONE receiver + TWO servos

- Configuration two: a wheel type transceiver + ONE receiver + ONE servo + ONE electric speed controller

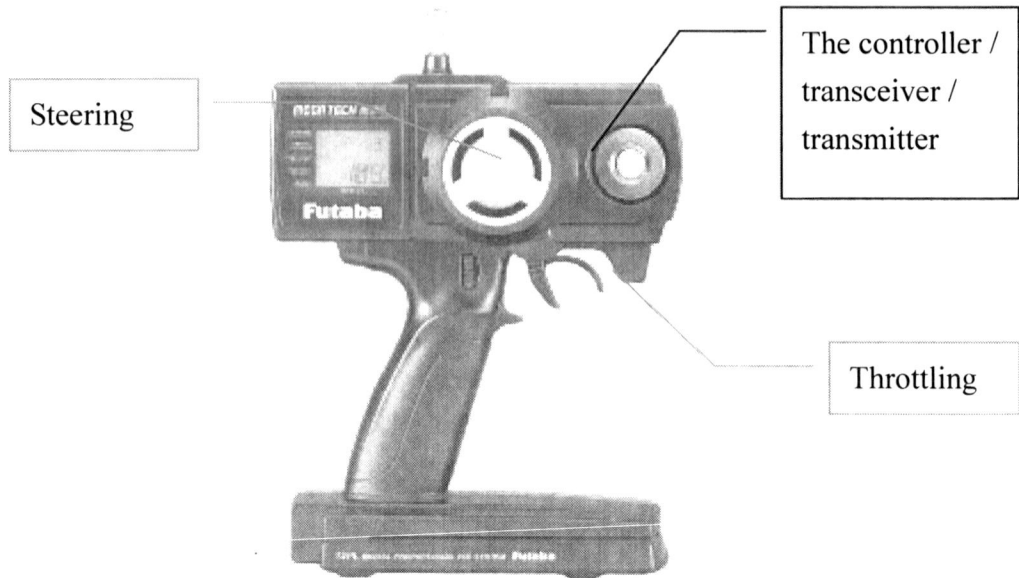

Steering

The controller / transceiver / transmitter

Throttling

The joystick type transceiver like the one show below is less popular for R/C cars.

NOTE: The most famous Radio Control devices makers in Japan are Futaba and KO Propo. Other famous names include JR, Airtronics, and Hitec.

Receiver and servo

In R/C terminology, a receiver is an electronic device that receives radio signal from the controller and decodes the signal for controlling the servos.

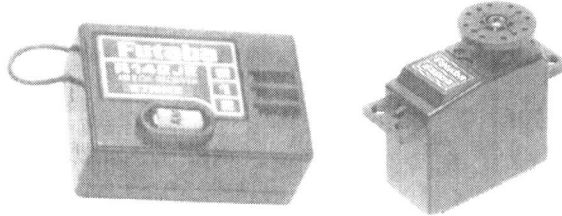

A servo (servomechanism), on the other hand, is a device for providing mechanical control remotely. R/C servos use an electric motor for creating mechanical force (and giving rotary output). R/C cars rely on servos for two purposes: for steering and for throttling (through controlling the mechanical speed controller).

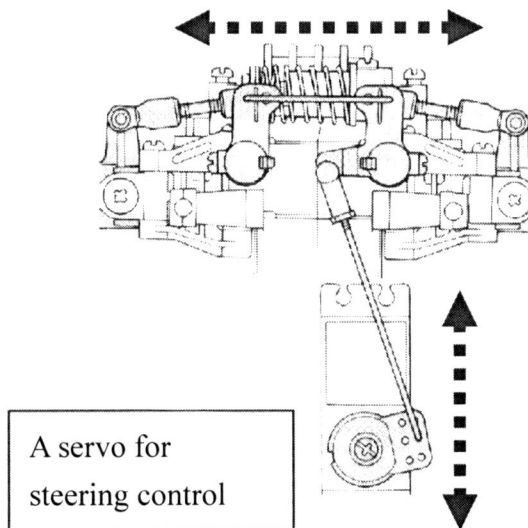

A servo for steering control

For steering, you want a servo with better response time and torque. You can check out the speed and torque from the servo specifications. For example, the Futaba BLS 452 servo offers a speed of 0.14 sec per 60 degree of movement, and with a torque of

14kgcm. The much cheaper S3003 offers a speed of 0.19 sec and a torque of 4.1kgcm. Note that some servos can work only with 4.8V power, while some can go for 6.0V. Digital servos are generally more expensive than the traditional analog servos.

On most EP cars, a servo saver is a piece of plastic device with a built-in spring for connecting the steering rods to the servo's output shaft. Its primary use is to absorb shock during driving or upon a crash. It is breakable, and in fact you want it to break upon a crash such that the servo itself can be free from any impact.

Servo saver

Some servos have built-in metal gears so they are less likely to break. Honestly, for normal racing it is very unlikely that you can break the servo UNLESS you are doing something extremely crazy.

Some crawlers have room for 4 wheel steering – two servos are required.

Servo with cored motor VERSUS Servo with coreless motor

Coreless motor and cored motor share the same design concept, with totally different assemblies. The in-depth tech details are out of the scope of this book. Generally speaking, coreless motors are of lighter weight and are capable of responding/accelerating/decelerating faster and smoother, with better precision and more power. And of course, they are more expensive than their cored counterparts.

Coreless analog servos and coreless digital servos are both available in the market.

Copyright 2010. **The R.C.PRESS (Hong Kong)**. All rights reserved.

Coreless servos are especially common for RC planes due to the lighter weight.

Analog servos VERSUS digital servos

They differ primarily in the servo amplifier, which is the device responsible for commanding the servo position. Analog device uses custom logic chip and timing components, while digital device uses Quartz crystal controlled microprocessor for processing receiver signals and controlling the motor. The microprocessor can always be fine tuned to adjust the amount of power sent for activating the motor, therefore allowing one to optimize the servo's performance basing on different track conditions. Also, as pulses may be sent to the motor at a much higher frequency, faster and smoother response as well as better resolution and holding power can be expected.

The performance gain is built at the expense of power consumption. Since the servo would have to draw power from the car's battery, you want to make sure you are using a high capacity battery. The traditional 1200mah battery pack may simply be insufficient for it. Keep in mind, digital servo would work only with digital receiver. Analog servo would only work with analog receiver. You just can't mix and match them.

FYI: A very obvious advantage of deploying 2.4Ghz radio is that you no longer need a very long "antenna" on the car. Also, there are way more available channels (there are about 80 channels in the 2.4GHz band) for you to choose from. In fact, some higher end transmitters will automatically select and lock onto clear channels for you.

Mixing and matching controllers, receivers and servos

You need to know that RC receivers are specific to a given frequency. The very basic requirement is that the receiver must match the frequency of the radio controller in order to be able to listen to its instructions. Within a frequency channel there are a number of sub frequencies for multiple racers to drive at any given time. If you are buying a RTR package the matching is already done for you. If you are buying these devices separately then you must manually ensure the crystal that is shipped with your receiver is matching the channel of your radio.

Modern RC radios are either AM or FM based. The crystal is either AM or FM specific and you can tell from the label attached to it.

An AM crystal on the sending end

FM radios may be grouped by those using positive shift and those using negative shift. Shift is about how the radio codes instructions for the receiver. It is very important that your receiver and your controller are using the same shift. Some come with the shift select/reverse shift feature for enhanced compatibility, while some don't. You just have to check the product specifications to clarify this.

For receiver and servo, you want to ensure the wire plug is compatible. Different brands have different wiring scheme for the plug, so compatibility is not always guaranteed.

Wire plugs

An AM crystal on the receiver end

PCM and PPM would further define how the radio controller codes commands that are to be sent to the receiver. Most AM radio use PPM. FM receivers, on the other hand, may be either PCM or PPM. As long as the shift matches, you should be able to mix brands of FM/PPM controllers and FM/PPM receivers without much problem. You can't do the same with PCM receivers as they are known to be quite brand-specific.

A Futaba AM based receiver

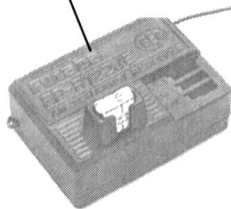

A Futaba PCM based receiver

Speed Controller: MSC versus ESC

Modern chassis designs do not have room for accommodating MSC.

With a MSC (mechanical speed controller), a resistor is used to limit the power passed to the motor. That is, the resistor is in action when you are reversing or driving at low speed (due to the need for limiting the current flow). This resistor can get overheated

Copyright 2010.

quite easily, especially when a high power motor is in use.

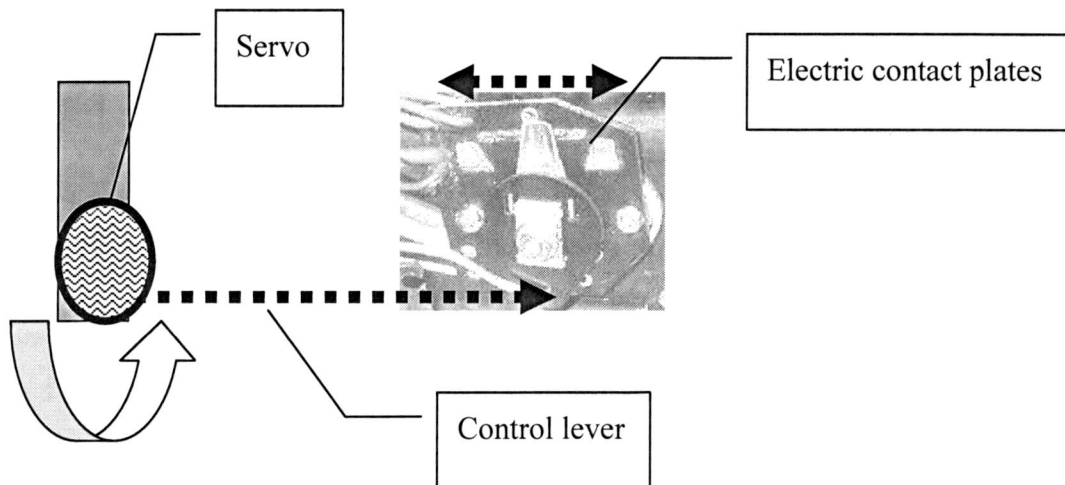

Servo

Electric contact plates

Control lever

If you go brushless then a ESC is a MUST. MSC does not support brushless operation.

An electric speed controller (ESC) is a popular alternative to the servo + mechanical

Copyright 2010.   **The R.C.PRESS (Hong Kong)**. All rights reserved.

speed control combination. It is preferable over the mechanical alternative because:

- weight-wise: the servo + mechanical controller combination is way heavier than a good ESC
- space-wise: the servo + mechanical controller combination occupies more space than an ESC
- performance-wise: an ESC gives longer run time, more top speed and allows for quicker starts as well as smoother running. It also gives room for finer adjustment of starting speed, top speed and braking power.
- reliability-wise: mechanical controller can easily burn when high current draw is passing through (in fact, sparks can kill the unit quite easily). In contrast, a good ESC is relatively more reliable and durable.

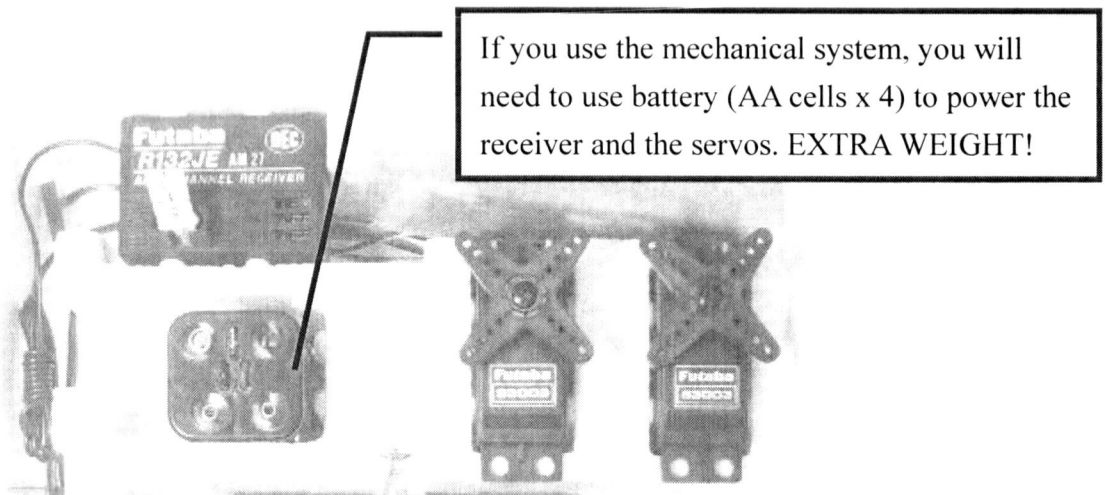

If you use the mechanical system, you will need to use battery (AA cells x 4) to power the receiver and the servos. EXTRA WEIGHT!

NOTE: ***Always go for an ESC. ESC is the preferred choice for modern R/C players. The mechanical speed controller is a thing of the past.***

Copyright 2010. **The R.C.PRESS (Hong Kong)**. All rights reserved.

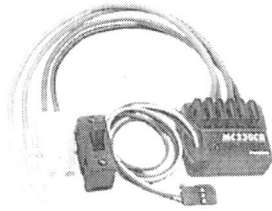

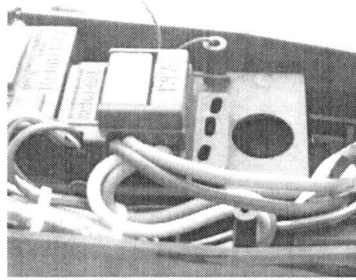

An ESC can be a stand-alone unit (most racing grade ESCs are standalone) which get plugged into the receiver's throttle control channel or incorporated into the receiver unit (as is the case in most toy-grade products, such as the QD series). All modern ESCs have battery eliminator circuit built in for regulating voltage (5V) for the receiver so that the need for extra batteries can be eliminated.

> NOTE: Generally speaking, ESCs are rated according to maximum current. The higher the rating the better if you are about to race. Do note that most racing grade ESCs do not offer the capability to reverse throttle.

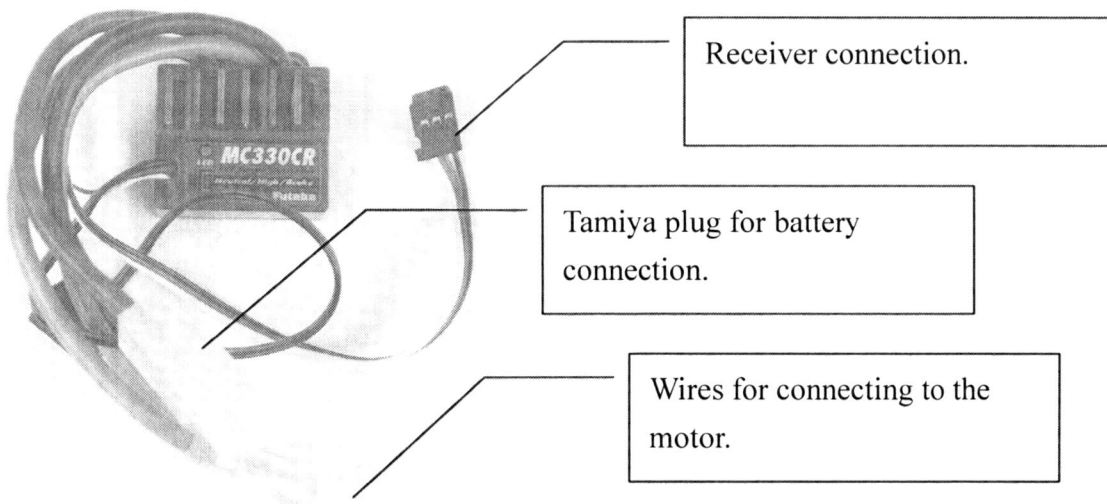

Receiver connection.

Tamiya plug for battery connection.

Wires for connecting to the motor.

The Twister 2.2 Carbon ESC offers a stunning max peak current of 3120A and a max continuous current of 780A. 780A is considered plentiful – more than sufficient for most racing conditions.

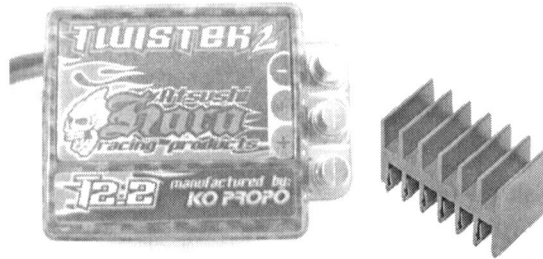

Heat is the primary source of harm for ESC. An ESC can get hot quickly if the motor is over geared or if the driveline is bound up. With the help of a heatsink the risk of overheating an ESC can be minimized. Seriously, you NEED the help of a heatsink almost all the time. Keep in mind, you need a heatsink that will physically match your ESC.   If the heatsink cannot be tightened on the ESC there will be no use at all. Not all ESCs are heatsink friendly though. Check with your dealer before buying one. **BL ESC would also require the use of cooling fan.**

Some ESCs provide fail-safe powers down on radio signal loss while some don't. You may want to figure this out from your vendor.

Cooking your ESC...

Using an over powered motor (a 18-T motor on a 20-T ESC) can cook your ESC almost immediately. Using an over powered battery (a 9.6V battery on a 7.2V ESC) can achieve the same. If there is significant binding in the drive train somewhere along the line or if you are running under heavy sunshine, you can still cook up your ESC in a minute or two.

Always allow for sufficient air inflow to your ESC. Don't shield it unless straightly necessary. Add a bigger heatsink. Check the gear ratio. To allow your motor to work under less stress you may want a slightly lower gear ratio. And check to make sure noting gets stuck along the drive train.

What if they got wet...

If your electronic components get wet....

The electric circuits won't get killed just by getting wet. It will get killed when it is wet and you attempt to use it - short circuit will result when electric current passes through.

If the electronic components get wet, the first thing you want to do is to unplug the battery. Take them out, dry the outside surfaces with a towel, then use a strong air blower or hair dryer (without heat) to dry them from every angle. Keep them idle for at least 24 hours before putting them back onto the car.

Mixing and Matching Receiver, ESC and Servo

To connect receiver and ESC/servo produced by different manufacturers, you may need to do some work on the connector. Say, if the connector plug has a shape that cannot fit into the receiver port BUT the wires seem to match (based on their colors, for example – White, Red, Black), one dirty trick is to open up the plastic shell of the receiver and file away the plastic that blocks the plug.

The two plugs have different shapes.

Copyright 2010.

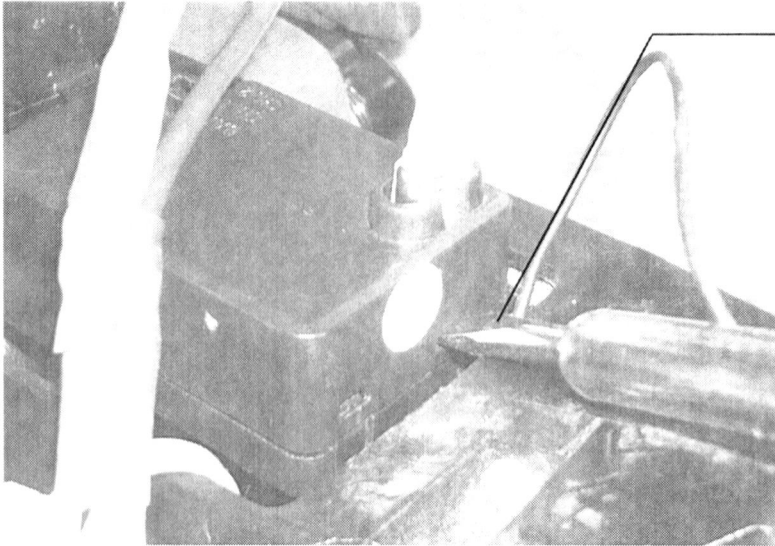

You can ply open the receiver shell.

Once the shell is opened you may do a little bit of filing to enlarge the connector port opening.

If the wire colorings are different, REMEMBER there is always one sure fire way to cook the receiver – that is, by having the RED wire in the wrong place. ALMOST ALL coloring schemes include a RED wire, and that RED wire carries power. If you put that in the wrong contact pin your receiver will hardly survive.

Expert tips

When shopping for a 2-Ch Radio Control System, pay attention to the following:

- Not all 2-Ch Radio Control Systems share the same wiring scheme. Therefore, if you are mixing and matching components of different brands, find out the schemes they use so you can properly connect them together. FYI, below is a table that summarizes the most popular wiring schemes:

"S" Type Universal Connector	Futaba "J" Connector	Wire Info
Brown Wire	Black Wire	Battery Negative
Red Wire	Red Wire	Battery Positive
Orange Wire	White Wire	Signal

- Different servos offer different torques. You want to have higher torque for off road buggies with heavy wheels (such as the Wild Willy).

- You need high torque when steering on land surface with extreme traction.

- Digital servo has built-in digital amplifier that features a high frequency pulse rate (sometimes as high as 300Hz) versus the standard 50Hz pulse rate found in most conventional servos, thus offering quicker and more accurate response. Unfortunately, digital servo packages are usually very expensive. Frankly, you don't really need it unless you are a real pro who needs absolutely unmatched response time on the track.

- Using the servo arm to turn the servo motor may cause servo gears breakage (most standard servo gears are plastic, although steel gears that are less strippable are available in certain high price models). Avoid doing this.

- Battery low on either side (the transmitter or the receiver) can lead to poor range or render your R/C car inoperable. **Always use good quality alkaline cells on the transmitter.**

- If your car does not steer even when battery is not low on both sides, check whether the connection between the servo and the receiver remains intact. If the connection has not been terminated, try to use another servo to test and find out if this is a receiver problem or a servo problem.

When shopping for an ESC, pay attention to the following:

1. What voltage level(s) can it support? Not all ESCs are 8.4V friendly.
2. Is the system a reversible one (meaning you can control your car to drive backward)?
3. What is the max current draw it can sustain (for forward / backward movement)?

Many ESCs are manufactured in Japan with specifications written in Japanese only. Below shows the specification of the Futaba MC230CR ESC:

電動カー用超小型FETアンプ（バック付）

使用電源	7.2V～8.4V（ニッカド6～7本パック）
ケース寸法	27.1×33.3×12.8mm（突起部を除く）
重量	44g
FET定格	
連続最大電流（前進）	90A
連続最大電流（後進）	45A
対応モーター	20T

FET speed controller for EP car(w/Reverse)
〈005275〉

Pay close attention to the lines highlighted:

- The first line says this ESC can take at the max 8.4V battery power input. Don't attempt to violate this. If it says 8.4V, then don't use 9.6V on it.
- The second one says it can sustain a current pass through of max 90A when accelerating forward (the higher the better. Personally I would go for one which can sustain at least 150A). **It basically tells the ESC tolerance level on the peak current your motor is to pull at full throttle when going forward.**
- The third one says it can sustain a current pass through of max 45A when driving backward (which means the system is a reversible one). Frankly this value does not really matter because the need for reverse throttling is rare – we usually need to do this only after a crash. And if you have a One-Way installed, reversibility would be totally meaningless.
- Then there is the line saying 20T. That means your motor should not go below 20T.

> NOTE: Both Noval and Tekin are very famous for producing highly reliable ESCs for the brushed community. In the brushless world, Castle Creations is a real big name.

The wires

The stock wires found in most R/C cars are of very high quality in terms of shielding – they are very difficult to tear. However, they are also very thin, meaning their ability to dissipate heat (heat generated by current flow) is not as good. Always remember, the larger the wire cross section the better able it is to dissipate heat. The problem with excessive heat is that the wire shielding will eventually get melted down. Our experience shows that most stock wires are good enough under relatively small mah battery (voltage is not the primary concern while the current is). If you are using high mah battery (such as a 3000mah one) to drive your R/C car, consider replacing all the stock wires.

When selecting the proper wire, pay attention to the American Wire Gauge (AWG) system. An AWG gauge of 14 or 15 can sustain higher amps for power transmission than the standard 16 or 18 wires.

12AWG Wire - good for up to 100+ amps. <- anything below 21-T
14AWG Wire - good for up to 50-75+ amps. <- anything between 21-T and 23T
16AWG Wire - good for up to 25-74 amps. ← **good enough for most stock R/C configurations running with RS-540**
18AWG Wire - good for up to 25 amps. ← in use by many cheap out-of-the-box configurations, NOT recommended at all

All wires have inherent resistance and that the best way to minimize resistance is to minimize the wire length (resistance is measured by having the resistance per unit of distance multiplied by the distance you are using). To do so you will have to trade off resistance and load with size and flexibility.

Choosing wire: Stranded wire VS solid wire

Generally speaking, there should be no less or more losses whether you use stranded or solid wire. One primary reason you wanna use stranded wire is ease of use - it is much easier to work with as it is very pliable.

On the other hand, frequent flexing and bending of solid wire can fatigue the wire and cause it to break. However, solid wire does have higher current carrying capacity in a smaller area (stranded wire is usually larger in diameter compared to the same gauge of solid wire - gauge is a measure of the total cross-sectional area of the conductor).

Note that stereo wires (speaker wires) were not designed to sustain high current draw and therefore should be avoided on your R/C car project.

The connectors

The Tamiya connectors are very popular among the standard battery packs. Since they can afford at the max 15 amp, they may not work for high power configurations. We have seen these connectors being melted down due to excessive heat. If you want connectors that can sustain high current, use Deans connectors instead:

With Deans connectors, the Female side is usually connected to the battery pack. Deans not only produce less resistant, but are physically more reliable too. Some say the Tamiya connectors have a very limited cycle life (approx. 50 cycles) before they worn out.

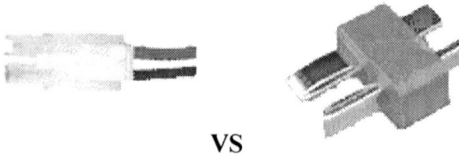

VS

If you insist on using the Tamiya connectors, note the max wire size restriction. Large Tamiya connector's max wire size is 14AWG, while small Tamiya connector's is 18AWG. Also, make sure you have wiring done correctly:

Large Tamiya Male: Red wire on the right (with the side of the clamp facing upward).

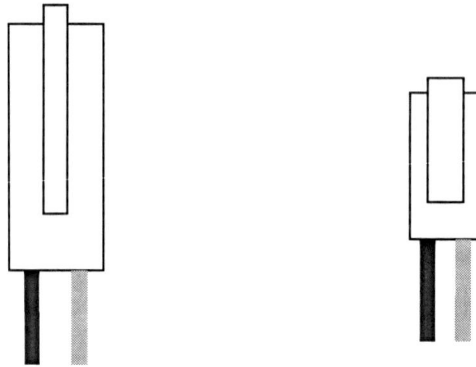

Small Tamiya Male: Red wire on the left (with the side of the clamp facing upward).

In any case, always keep these in mind:
- Use red wire for all connections to positive (+) voltage.
- Use black wire for all connections to negative (-) voltage.

You will need to make sure the correct type of plugs are being used with the correct type of connector. The connector attached to the battery has no clamp on it and is using male plugs. The connector of the ESC/charger has a clamp and is using female plugs.

If your battery charger has a small Tamiya connector, you may create your own converter cable (do remember to match pos and neg correctly or you will damage your battery):

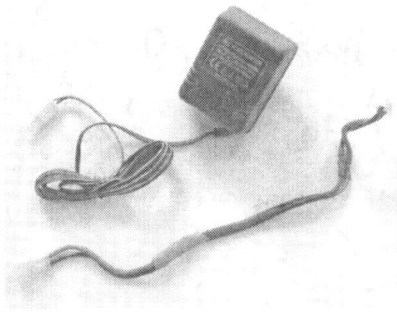

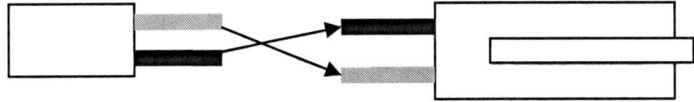

Upon frequent plug/unplug processes the plugs can eventually go loose. You may want to replace the plugs and the connectors altogether (they are cheap anyway) for, say, every 40~50 rounds of race.

The batteries

Nowadays the trend for 1/10 R/C car racing is to go with 7.2V, although occasionally you may find tracks which allow the use of 8.4V. 9.6V is seldom allowed (in fact, very few ESCs in the market support 9.6V).

The quality of your battery pack is very important. In fact, with a good battery even a stock RS540 motor can run at its optimal efficiency.

Copyright 2010. **The R.C.PRESS (Hong Kong)**.

Discharge characteristics

Different types of battery have different discharge characteristics. Refer to the graphs below, batteries of the rapid discharge type (such as the Sanyo SCR cells) can discharge faster, thus giving a more solid punch to the motor at the expense of endurance.

Batteries with a rapid discharge characteristic tend to have a sudden and abrupt drop of output voltage (to a point which can no longer turn the motor):

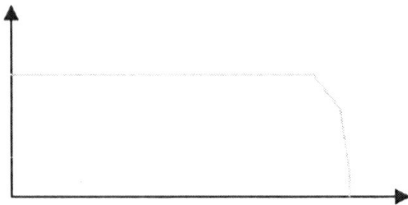

On the other hand, batteries with a slow discharge characteristic tend to drop its output slowly and gradually (which means a period of slow speed driving is possible before getting totally idle):

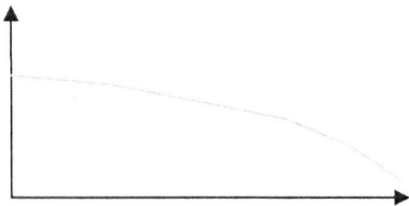

Voltage VS mah

Many people prefer to classify batteries based on voltage and mah. They like to look into these two factors separately. Without going into the complicated dynamics of battery and electric motor, we wanna tell you that battery performance is a function of both voltage AND mah COMBINED. When we say high voltage is dangerous for the speed controller, we usually mean large full size battery pack which has both high voltage and high mah output.

Although mah and resistance are not necessarily linked, large batteries tend to have higher mah value and at the same time less internal resistance, meaning they can sustain higher current draw at any given time. Simply put, higher mah means higher capacity AND ALSO the ability to sustain current draw at a higher rate. Effectively the motor can demand more from the battery at any given moment.

NOTE:	For those of you who are interested in learning the basics of electricity:
	In the world of electrical circuit, the number of electrons that are moving is called the current, and the pressure pushing the electrons along is called the voltage. Resistant is the measurement of the ability of electrons to move through the pipeline.
	The three most basic units in electricity are voltage (V), current (I) and resistance (r). Voltage is measured in volts, current is measured in amps and resistance is measured in ohms. The basic equation in electrical engineering says that the current is equal to the voltage divided by the resistance: I = V/r
	Try to think of how water flows through a pipe:
	• Increasing water pressure causes more water to flow through the pipe [higher voltage].
	• Increasing the size of the pipe allows more water to flow through the pipe at any given time [higher capacity and less resistant].
	• Both of the above factors can increase the supply of water.

However, if you have a large pipe but insufficient water, the water supply won't increase. On the other hand, if you have sufficient water but a pipe that is too small, a bottleneck is in effect.

- In theory, increasing either the current or the voltage can result in higher power. However, there is an advantage that comes from using less current to make the same amount of power. Since the resistance in electrical wires consumes power, the power consumed will increase as the current going through the wires increases. By using a higher voltage to reduce the current, one can make the electrical system more efficient.

Using small batteries

Smaller batteries have higher internal resistance and can get heated up easily. If you really want to use smaller batteries for fitting into the chassis, you may need to raise the voltage. Practically speaking, the performance output of a 8.4V Sanyo 600mah is roughly on par with that of a 7.2V GP 2000 mah (endurance is a different story though). And 600mah is way too insufficient. If you really need to go small, consider the GP 1100mah NiMHs instead.

As said before, the small batteries are not really sufficient for powering an upgraded car. If free space is available but is "dispersed" all over the chassis, you might consider using a parallel configuration. That is, you can have multiple sets of small batteries chaining up together to double the capacity. Let's refer to this example:

- If you arrange four 1.25-volt, 500 milliamp-hour batteries in a serial arrangement, you get 5 volts (1.25 x 4) at 500 milliamp-hours.
- If you arrange four 1.25-volt, 500 milliamp-hour batteries in parallel, you get

1.25 volts at 2,000 (500 x 4) milliamp-hours.

- Do you know that a standard 9V battery is in fact created by assembling 6 x 1.5V batteries together?

With a serial configuration, you get higher voltage. With a parallel one, you get higher current for sustaining higher mah. Do note that the connection must be parallel, NOT serial:

Parallel Connection
(doubled capacity)

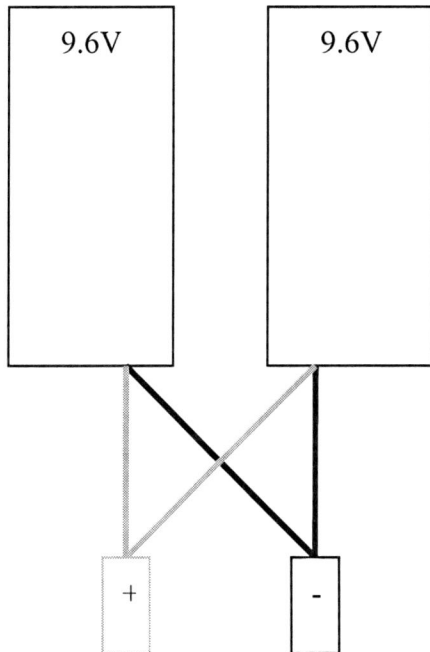

Serial Connection
(doubled voltage)

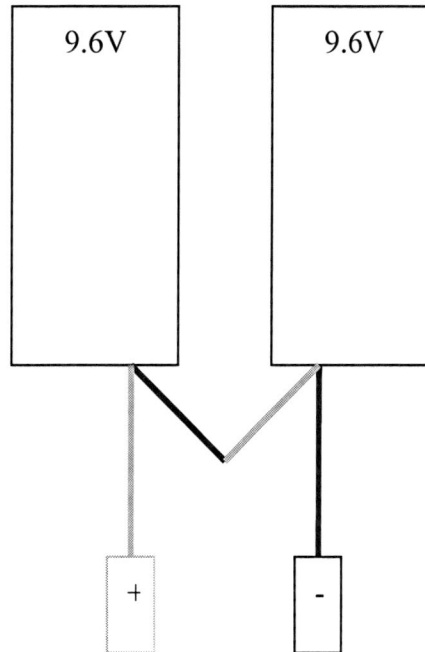

9.6V	9.6V

+   -

9.6V	9.6V

+   -

Wiring can become complicated, but the power of two joining forces together can produce the horse power required by an upgraded motor. Do note that once they are connected together, you should never have them charged and discharged separately. Always treat them as a single entity.

> NOTE: When pairing two battery packs, it's recommended that you use packs of the same size, make, and capacity. They should be the same as to having the same charging/discharging characteristics so that one pack does not get overcharged while the other is only half charged. If this happens you will kill your pack much much faster. It is also recommended that you assemble the packs when they are both at a correctly discharged state, then charge/discharge as one unit. If you build a parallel battery pack from one charged pack and one discharged pack, one will draw current from the other, which can be harmful for both packs.

NiCds VS NiMHs

Ni-Cds have good performance in high-discharge and low-temperature applications, but are subject to the problem of "memory effect" (the cell remembers and maintains the characteristics of the previous cycle). They will also self-discharge at a rate of about 1% a day. Sanyo and Panasonic offer good quality NiCds.

NiMHs closely resemble the NiCds in construction but can offer 30% more capacity and about 25% more power than the NiCd counterparts. The problem is that they have poorer service life, tend to self discharge faster, and in certain cases are subject to "memory effect".

Sanyo, Panasonic and GP

Based on our experience, Sanyo and Panasonic have good Nicds, while GP has good NiMHs. Be very careful when buying Sanyo cells, as there are many fake Sanyo cells in the market nowadays. Frankly, it is not easy to tell from the outside whether a Sanyo cell is authentic or not.

Prolonging the life of the battery

Heat is the largest enemy of your battery as it causes the separator and seals to weaken and accelerates changes in the plate material, which in turn causes the dreaded memory effect.

Overcharging abuse is one easy way to overheat your battery. In theory, most battery cells have incorporated a protection mechanism against overcharging through increasing their ability to resist venting. However, heat is still generated along the process and is causing irreversible damage.

To avoid overcharging, use computerized charger (those capable of pulse charging) if you have a deep pocket, or use slow charging if you are not in a rush. Slow charging is always safe since it is too slow to overblow the battery.

If you really need to get the battery charged in less than an hour, set an alarm to remind yourself when to stop the charger, or use a failsafe timer if one is handy.

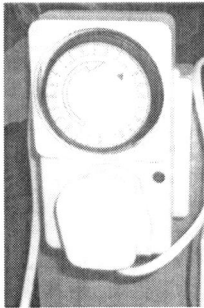

A low cost mechanical timer unit

In any case, it would be a good idea to have a fan blowing at the battery for keeping it cool.

The charger

The most basic type of battery charger is the "constant voltage, current limited" type, which can be made with a transformer (which converts house voltage to battery voltage), a rectifier (which converts the household alternating current (AC) to battery direct current (DC)), and a resistor (which limits the current as a protection for both the charger and the battery). These chargers are cheap and are very good at providing a normal bulk charge but are never good at finishing a charge or maintaining a battery.

Low cost wall charger (slow)   Low cost fast charger (no timer)

Smart Chargers using computer technology can perform step regulated charging techniques. They are expensive (some real good ones cost well over USD$100), but are capable of tailoring charging current for rapid restoration of battery capacity due to the incorporation of controls for separating battery charging into finer stages.

Microprocessor controlled charger like the DuraTrax Piranha AC/DC Digital Peak Charger offers six programmable features, including adjustable charge current and peak detection sensitivity (http://www.duratrax.com/caraccys/dtxp4005.html). If you want to learn advanced battery charging, here is the URL of a MUST VISIT page: http://www.angelfire.com/electronic/hayles/charge1.html .

Always remember, batteries like to be charged in a certain way. Ask your dealer for this kind of information when buying a battery pack.

Automatic Peak Charger/Discharger:

Do whatever you can to keep the charger cool. I add a large heatsink to my charger (as long as both sides have flat metal surface then any heatsink will do):

The kind of thermal grease used by computer CPU heatsink was applied, which helped in sticking together the heatsink and the charger shell.

Slow charging VS fast charging

NOTE: *Different types of battery have different charge & discharge characteristics. Not all types of battery are suitable for fast charging.*

One major reason why slow charging is good is because it is less likely to overheat the battery. However, very slow charging (those over a day) may also not be as good.

Under moderate charging currents, the cadmium that is deposited inside the battery pack are very small crystals. Given time, these crystals tend to coalesce and form larger crystals. This is no good for the battery since it makes the cadmium harder to dissolve during high current discharge (the battery is in high current discharge stage when the R/C car is running at full speed on the track), and can lead to high internal resistance as well as voltage depression. Very slow charging is no good because slow growth aids the formation of larger crystals.

Based on our experience, getting a battery pack fully charged in about 5~6 hours is the most ideal. One common way to judge whether a battery is fully charged is through calculating the output of the charger to match the capacity of the battery pack. Another cheap way is to measure the battery temperature by hand (assuming normal room temperature). If you feel that the battery temperature is rising steeply, full charge is almost reached.

Memory effect

With memory effect taking place, the battery retains the characteristics of the previous charge cycle. This can happen when the battery is recharged without being fully discharged, and can result in temporary loss of capacity. The memory effect was once thought to be absent from NiMH battery. However, it is now being recognized that memory effect can occur on NiMH battery. Refer to the next section for information on how to avoid memory effect.

To prevent the problem from happening: You do NOT need to cycle your battery each time it is used. In fact, DON'T deliberately discharge the battery to avoid memory effect as this will shorten its life span. A good way to avoid memory effect is to use the battery until it is about 90%~95% discharged. Do this only once in a while. And, fully discharge the battery once every 30 or so cycles. **To correct the problem:** Simply perform a few cycles of slow discharging and slow charging (but don't go too slow – charging and discharging over a day is no good as well).

Proper charger configuration

We are using the 3Racing C6 Pro charger for demonstration here. The C6 Pro is a microprocessor controlled fast charger which can charge NiCD, NiMH, LiPo and LiFe packs. When it is microprocessor controlled that means there are settings you can manipulate through the buttons and the LED display.

Just like other micro processor controlled unit, the C6 Pro itself does not come with a power supply. That means you need to have your own.

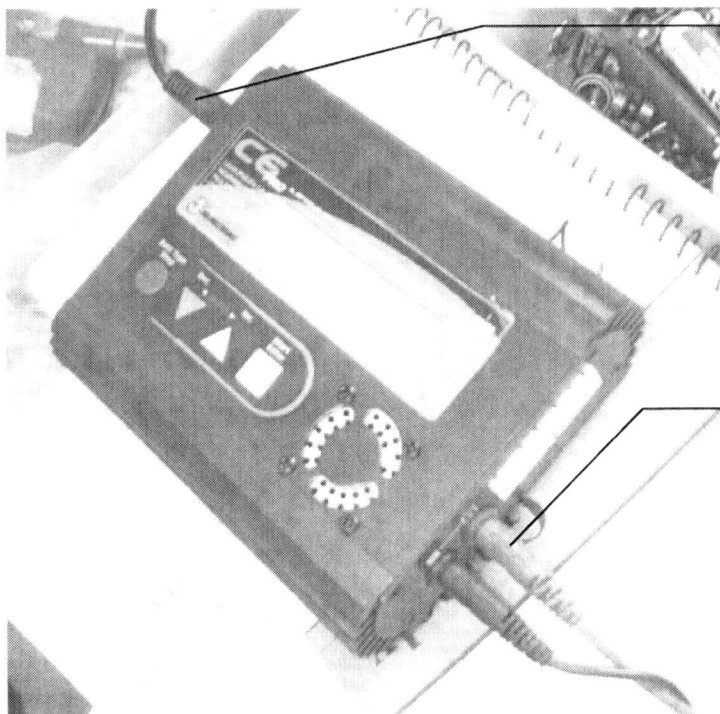

Input power connection, which is DC only. You may use a separate power supply unit or a car battery as the power source.

Output

Copyright 2010. **The R.C.PRESS (Hong Kong)**. All rights reserved.

The trick is that since the C6 Pro has a max charge current of 5.0A, you will need to use a power supply which can sustain a current draw which is 2X or 3X of this amount. Anything less and your charging configuration will simply not work in a stable manner. Those small wall mount power supply with 1A or less will refuse to work even if you adjust the charge current to the lowest limit of 0.1A on the C6 Pro.

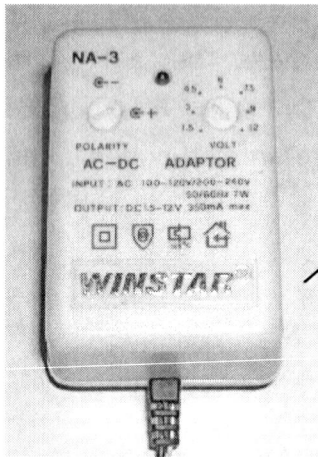

DON'T USE THIS KIND OF POWER SUPPLY. IT IS WAY TOO WEAK.

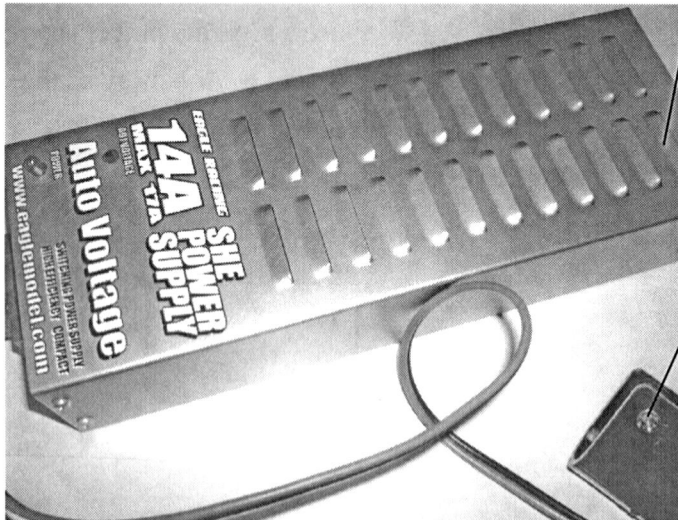

This large size SHE Power 14A power supply is good enough for providing stable current to the charger.

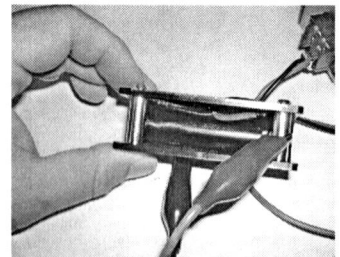

Copyright 2010. **The R.C.PRESS (Hong Kong).**

And by the way, the stronger the current support the faster you can complete the charging process. Say if you run the C6 Pro at its max limit of 5A (5000ma per hour), a 1700mah pack can be fully charged in about 1/3 hour. If you run it at 0.1A. charging would take more than a day to complete.

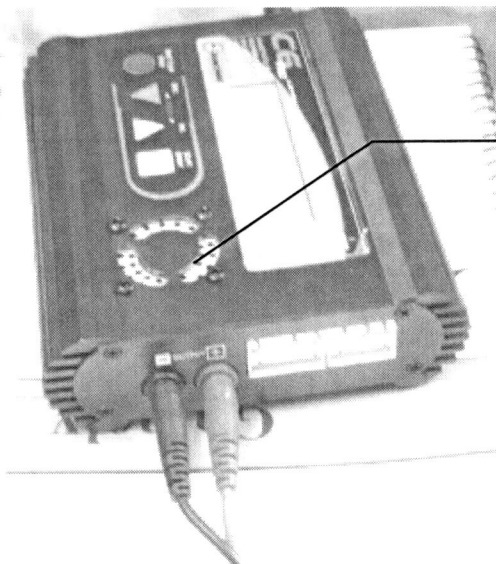

Do NOT cover this opening as there is a cooling fan underneath it for taking out heat from inside the unit.

When you connect the output, make sure you don't mix up the RED and BLACK wires. The connection ports on the right are for balancing of Lipo pack. To perform balanced charging BOTH SIDES would need to be connected.

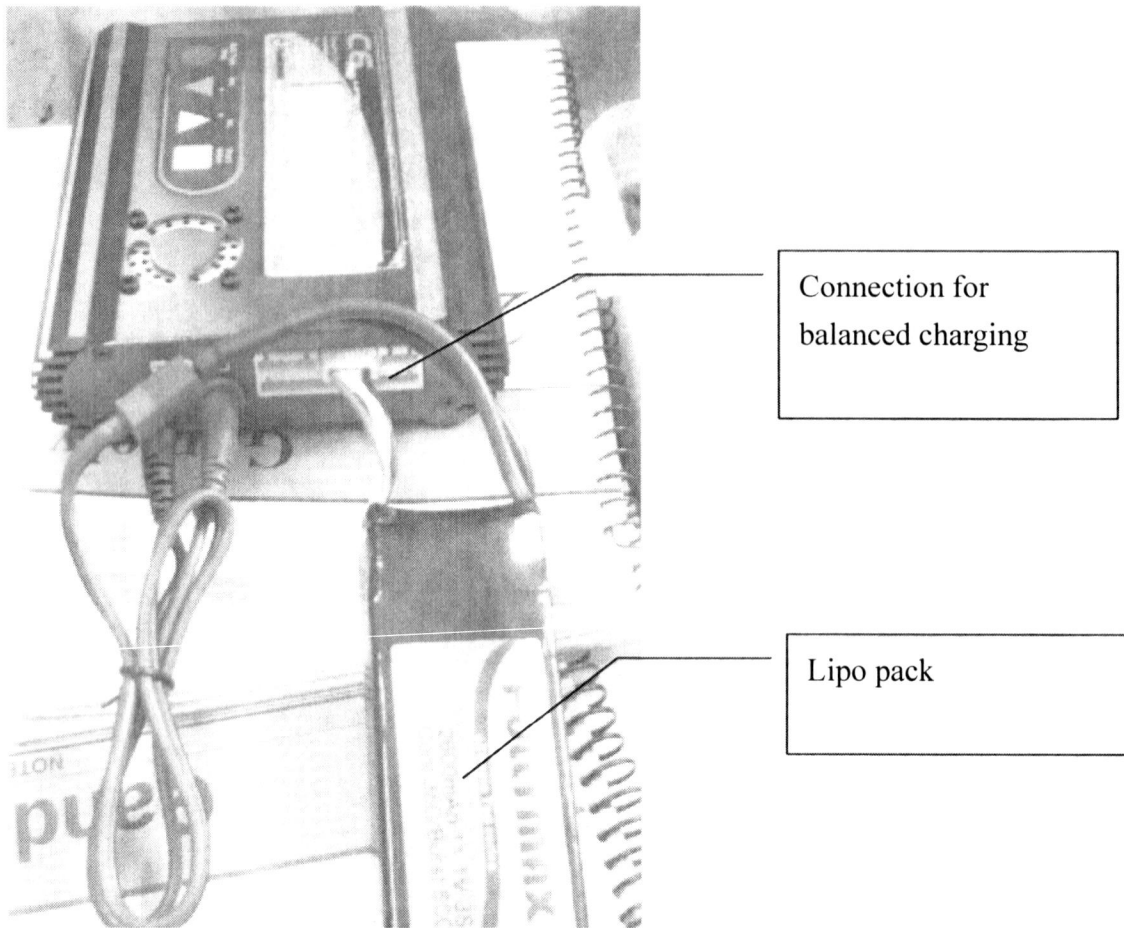

Connection for balanced charging

Lipo pack

And keep in mind, even though microprocessor controlled chargers like the C6 Pro are all said to be capable of auto detecting the battery status, it is still advisable fort you to manually verify the charging configuration every time a different battery pack is being charged. And never leave the charging process unattended, especially when you are dealing with Lipo packs.

SPEED has a similar charging option to offer. The EX-1506 is fully computerized

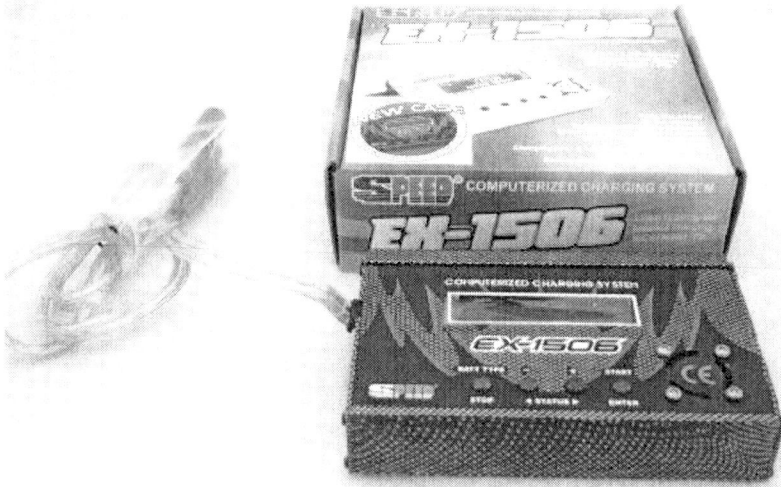

The EX- 2708 is similar in functionality but can support more cells (higher output voltage).

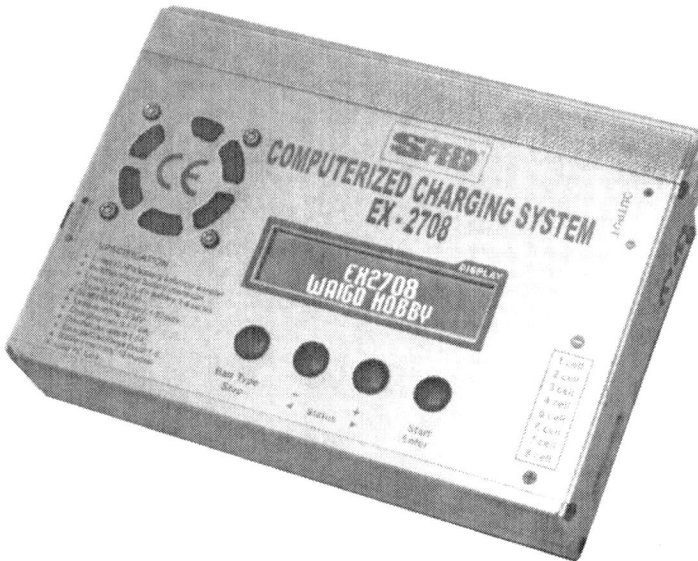

The MAX4QUAD Twin Dual Channel Charger is another very cool charger option. It can charge two battery packs at a time, and if you connect four to the charger, charging will

Copyright 2010.   **The R.C.PRESS (Hong Kong)**. All rights reserved.

take turns to get completed. One major difference between this charger and the EX1506 is that this charger can directly connect to home power, while the 1506 would require a separate power supply to operate.

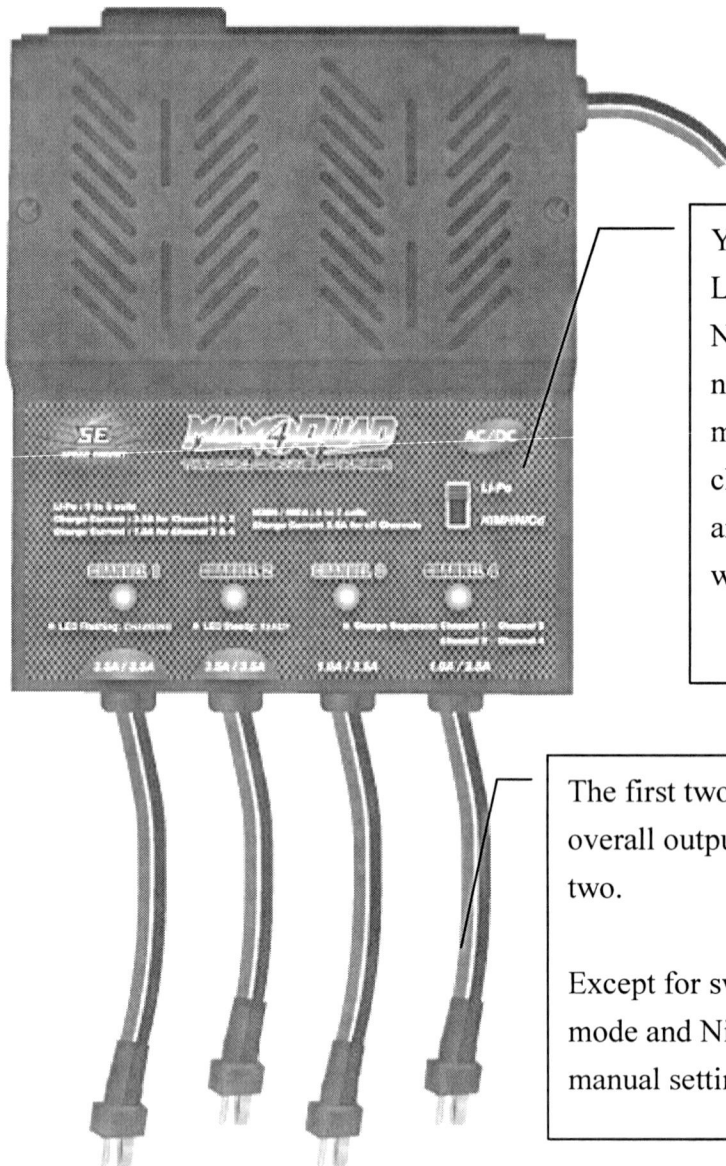

You can switch between Lipo mode and NIMH/NICD mode. Do note that when attaching multiple packs for charging you cannot mix and match Lipo pack with regular pack.

The first two output lines offer higher overall output power than the second two.

Except for switching between Lipo mode and NiMH/NiCD mode, no manual setting is necessary.

When charging multiple batteries together, make sure they are of the same type (Lipo or non-lipo). Mixing and matching batteries of different capacity (different mah values) is okay though.

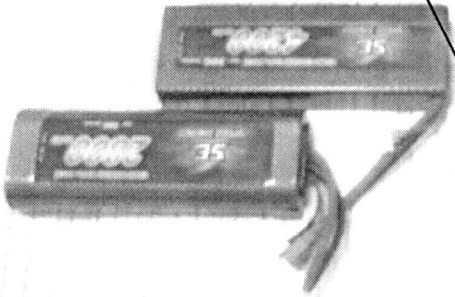

Balancer units x 2 are included for Lipo charging.

Expert tips

- If you are making your own custom pack, pay attention to the following points:
    - before soldering on any high capacity NiMH cells, have them fully discharged (as they usually come approx. 40% charged out of the box).
    - use soldering pen capable of 80W or above, complete soldering ASAP to avoid overheating the cells.
    - after soldering the cells together you should first wait till they cool off before charging.

- NEVER mix and match NiCDs and NiMHs!

- Those cheap and slow wall chargers will work safely with both NiCad and NiMH batteries. For fast computerized charging, you may want to use a charger which specifies that it is designed for NiMH. As the characteristics of NiCad and NiMH cells are slightly different, different programs may be necessary for terminating the charge in different ways.

- For peak performance, use your battery immediately after charging. Voltage usually starts to drop in 15 minutes after charging is completed.

- To avoid over-discharging (and hurting) the battery, withdraw your car from the track if the car starts to slow down 20% or so. Don't work on the battery until it gets cooled down completely.

- Give your quick charger a heatsink if it has a flat surface on its metal casing. A cooling fan blowing at the heatsink is of course preferred.

- Under extremely cold weather, you may want to warm up your battery (to room temperature of 20~25 degree Celsius) before charging. Battery performance can be negatively affected when it gets too cold.

- If you do not have the budget necessary for buying a computerized charger, consider buying a cheap mechanical timer for use with your low cost charger. Remember, overcharging a battery is a human error that occurs almost everyday. Do whatever you can to avoid this.

- A matched battery pack is one which has been processed through a computerized matching process such that all cells in the pack are close to the same quality. Most matched batteries come pre-built, although you can buy a package of matched cells and go the DIY route. Keep in mind:

- you should never mix and match matched cells of different grades.

- for maximum performance you need a real good charger and discharger for use with a matched pack. The whole setup can be expensive. As a beginner you don't really have to go this far...

Troubleshooting R/C electronics

Cleaning of electrical contacts

On a traditional mechanical speed controller you have electrical contacts around the "switch assembly". Proper maintenance of these contacts can extend their usable life. They don't last forever, but if you do not overload them and you keep them clean, there shouldn't be any need for frequent replacement. Wipe the metal surfaces clean. Use Contact Cleaner if necessary to clean the carbon deposits and other settlements on the electrical contacts. Ensure that no residue is left on the metal surfaces. If there are signs of corrosion, use a Dremel tool with brush to clean the contacts lightly.

Troubleshooting R/C electronics

> NOTE: Sometimes problem can occur simply because the metallic plugs inside the Tamiya connectors are getting loose.

In the context of R/C, you may use a multimeter to make various electrical measurements, such as DC voltage, DC current, and resistance. It is the best tool to use for troubleshooting R/C electronics. A multimeter is called a multimeter because it combines the functions of a voltmeter, ammeter, and ohmmeter (an ammeter measures current, a voltmeter measures the potential difference in voltage between two points, and an ohmmeter measures resistance. A multimeter combines these functions into a single instrument). Digital meters give an output in numbers, usually on a liquid crystal display. On the other hand, analogue meters show output by moving a needle along a scale.

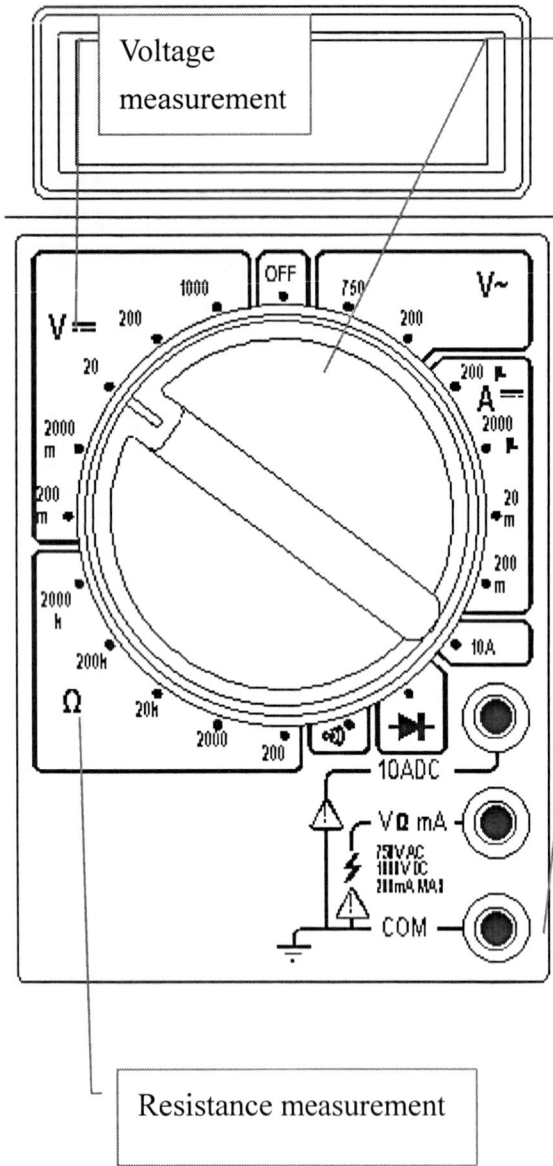

Voltage measurement

With the central knob you choose which voltage level is appropriate for the measurement you want to make. If the meter is switched to 10V DC, for example, then 10V is the maximum voltage you can measure.

The black lead is always plugged into the common terminal (the terminal marked COM). The red lead is plugged into the 10 A jack when measuring currents greater than 300 mA (this is the typical limit - check your multimeter for possible variation). When measuring currents less than 300 mA, you may need to use the remaining jack (again, check your multimeter specification). In fact, to avoid blowing the input fuse, use the 10 A jack until you are sure that the current is less than 300 mA.

Resistance measurement

Copyright 2010. **The R.C.PRESS**

What is "DC"? DC stands for direct current. In any circuit which operates from a steady voltage source (such as the battery used by our R/C cars), current flow is always in the same direction. Every R/C car works in this way.

You usually measure voltage to determine the health and status of your battery. If, let's say, a 9.6V battery has a voltage reading of less than 9V, your battery is pretty weak and may require a full recharge and re-test. If a weak reading is constantly achieved, battery replacement may be necessary.

You usually measure resistance to test for connectivity. If resistance exists but is less than a certain threshold as determined by the specs of your multimeter (a typical threshold value is 210 ohms), the beeper on the multimeter may sound continuously to indicate the existence of connectivity.

NOTE: You may also measure resistance to determine if improvement has been achieved after a major rewiring work. You get less resistance than before if there is an improvement. Do understand that resistance testing works by passing a small current through the component and measuring the voltage produced. Therefore, if you try this with the component connected into a circuit with a power supply (such as a battery), the meter's fuse may be blown.

Special Notes: Upgrading the QD (Quick Drive) R/C cars

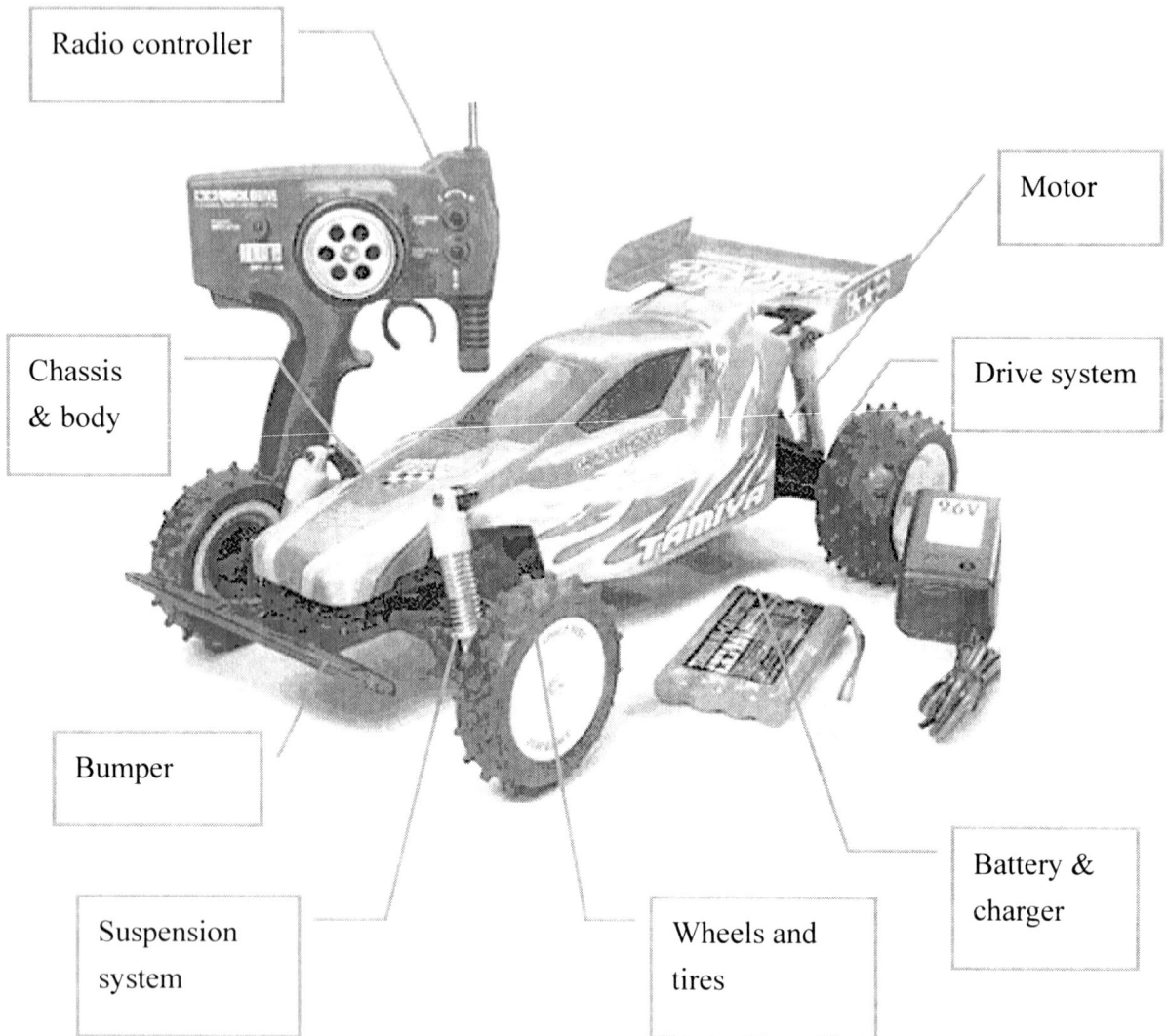

Radio controller

Motor

Chassis & body

Drive system

Bumper

Suspension system

Wheels and tires

Battery & charger

Upgrade objective

Due to the inherited limitations of the QD architecture, you can't expect much in terms of top speed improvement (a 10 ~ 15% speed improvement is considered reasonable for a QD car). You should focus more on the Controllability and Endurance aspects instead.

Chassis

Basically, on a typical ABS plastic chassis like the type used by a QD, you may drill holes on areas which do not affect the overall sturdiness of the chassis platform (of course you don't want a car full of holes on the chassis to run on dirty tracks).

Weight distribution is another issue. Due to the fact that most QD R/C cars have their motor placed closer to the backend of the chassis, the rear and front ride heights tend to be uneven (the heavier side goes lower). An artificially configured "lower front" can benefit steering but is easier to go "diving" (which can make the car slides) when the brake is hit hard.

QD dampers are purely spring based.

Due to the "speed limit" imposed by the 280 class motor, you can't really damage the chassis unless you intentionally drive the car towards the wall at full speed for a hundred times. Even if you drill holes on the chassis, the chassis shall not break that easily.

Battery compartment

The way your battery is housed depends primarily on the chassis layout. In any case, the battery compartments of most QD cars are not flexible enough to allow for anything larger than a standard 9.6V AA cells pack. If you want to use the larger full size cells, you will need to do lots of cutting and trimming plus creative wiring.

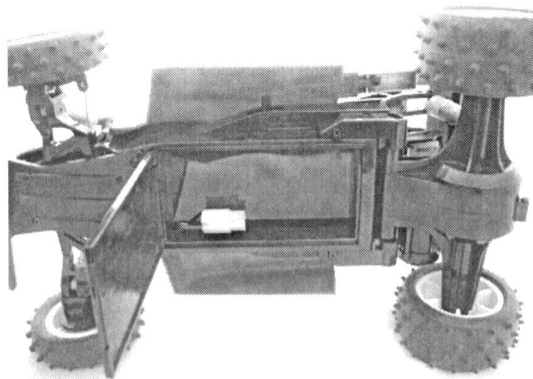

Some QD cars (or QD compatible cars) have a wide front bumper out of the box which can effectively protect the front arm assemblies. However, most other QD cars are equipped with small bumpers due to the need for fitting with the plastic body shell.

Gear switching

Almost all QD cars include a switch lever at the back of the gear box for manually switching between low gears and high gears. Going uphill or riding on grass land constantly with low torque gears (gears with a high ratio) can shorten the life of the motor significantly.

NOTE:	"Gear pitch" can be thought of as the closeness of the gear teeth. At present there is no gearbox parts available in the market for

The 280 motors

Most off-the-shelf QD R/C car kits from Japan are shipped with the Mabuchi 280 motor. Mabuchi 280 is a closed endbell motor which is well built and understressed for offering moderate performance balanced against durability. The diagram and table below show the dimensions and specifications of the Mabuchi 280 series motors. Note that they come in a number of variations:

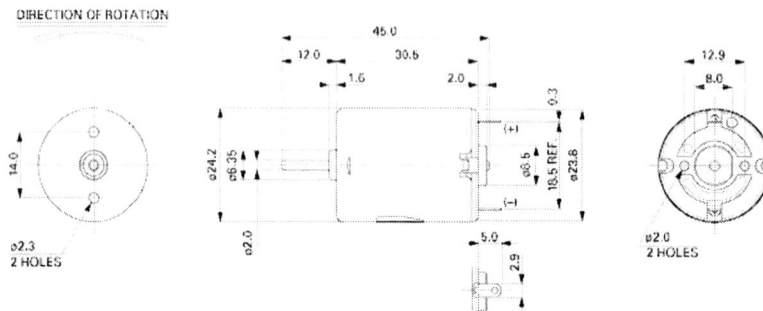

MODEL		VOLTAGE		NO LOAD		AT MAXIMUM EFFICIENCY				
		OPERATING RANGE	NOMINAL V	SPEED r/min	CURRENT A	SPEED r/min	CURRENT A	TORQUE mN·m	g·cm	OUTPUT W
RC-280RA	2865	4.5 - 6.0	4.5	13600	0.27	11350	1.36	2.92	29.8	3.47
	20120	4.5 - 9.0	6	9900	0.12	8210	0.58	2.26	23.0	1.94
	2485	4.5 - 8.4	7.2	16800	0.21	14260	1.18	3.27	33.3	4.87
RC-280SA	2865	4.5 - 9.0	6	14000	0.28	11910	1.60	4.46	45.5	5.58
	20120	4.5 - 12.0	12	15500	0.15	13320	0.92	4.68	47.7	6.52
	2485	4.5 - 9.6	6	10800	0.18	9130	0.99	3.71	37.8	3.54

NOTE: Johnson Electrics also produces 280 compatible motors for the R/C market. The specifications are slightly different though.

The more powerful 370 SD motor (which has a slightly different dimension) can be used for substituting the 280 (it offers higher torque than most 280s):

MODEL		VOLTAGE		NO LOAD		AT MAXIMUM EFFICIENCY			
		OPERATING RANGE	NOMINAL	SPEED	CURRENT	SPEED	CURRENT	TORQUE	OUTPUT
			V	r/min	A	r/min	A	mN·m	g·cm	W
PK-370SD	2870(*1)	4.5 - 9.6	7.2	16500	0.34	13790	1.73	5.97	60.9	8.61

> **NOTE:** The stock 280 motor has a closed endbell, meaning there is no way you can have it disassembled without breaking the motor. You may not want to risk doing this.

Motor compartment is small.

The batteries

Nowadays the trend for 1/10 QD cars is to go with 9.6V. The stock Tamiya QD battery is 600mah. For better performance (top speed and endurance), you may want to go for something with larger capacity, such as the 2100mah pack shown below:

NOTE: *The quality of your battery pack is very important. In fact, with a good battery pack even a stock 280 motor can run at its optimal efficiency and deliver 10%+ performance improvement.*

Copyright 2010. **The R.C.PRESS (Hong Kong)**. All rights reserved.

For the latest product releases, please visit:

www.rcpress.com

You may also contact us via this email address:

editor@rcpress.com

AirsoftPRESS publishes ebooks and printed books on airsoft technology.

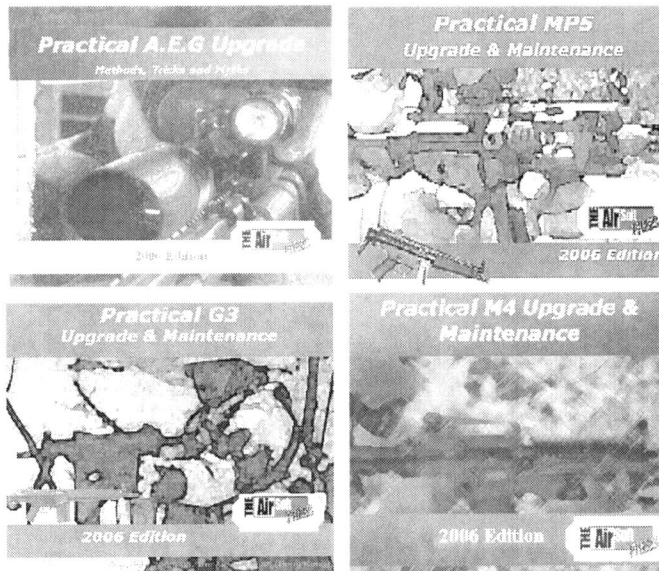

www.airsoftpress.com

Team TEKIN is the expert in brushless system:

www.teamtekin.com

Telebee produces affordable yet reliable brushed ESCs.

www.telebee.com.hk

3Racing offers high quality upgrade parts for most Tamiya EP car models:

Model No.: DF03-01/V2/LB
Name: Aluminum Oil Damper Set Ver. 2 For DF-03
Status: New Arrival

Model No.: DF03-01RF
Name: Rebuild Kit (...t) Fo... ...1/LB
Status: In Stock

Model No.:
Name:
S...

Model N...

**Visit
www.3racing.hk
for a full catalog.**

Model No.: ...avy Duty Ver. 2
Status: New Arrival

Model No.: DF03-05/LB
Name: Auminium Propeller Joint For DF-03
Status: Out of Stock

Model No.: DF03-06/WO
Name: Rear Graphite Shock Tower For DF-03
Status: In Stock

Model No.: DF03-07/LB
Name: Rear Aluminum Hub Carrier For DF-03
Status: Out of Stock

偉高模型
WAIGO HOBBY

● Address. G/F., No.7-8 Tung Fong Street, Yaumatei, Kowloon, HK
● Tel: 2384 0003   ● Fax: 2710 8861
● E-Mail: inquiry@waigohobby.com   ● Wedsite: www.waigohobby.com

OFFICAL DISTRIBUTED BY WAIGO MODEL HOBBIES LTD.

Hybrid heat radiation system "H.T.R.S." (Patent pending)

Brushless amplifier with the world's-lowest ON resistance

No PC is required.
Settings can be changed using the ESC buttons only. (Free operation system)

High-density 12-gauge cable comprising 1,530 wires

Smooth operation   Optimized KV setting

Use of heat-resistant sintered
neodymium magnet of the highest class in the industry

The sensor control system   Low resistance system
enables more accurate motor control.   to prevent current loss

**PEARL Evolution
Brushless ESC**
NOS 90900

**PURE Evolution
BL Modified Brushless Motor**
NOS 90647 9.5T 3800kV

**PURE Evo
SpecRacing Brushless Motor**
NOS 91831 10.5T 3600kV

CPSIA information can be obtained at www.ICGtesting.com
Printed in the USA
LVOW09s2006101115

461894LV00006B/566/P